my **revision** notes

OCR B (MEI) A Level

MATHEMATICS YEAR 1/AS (APPLIED)

Stella Dudzic
Rose Jewell

HODDER
EDUCATION
AN HACHETTE UK COMPANY

The Publishers would like to thank the following for permission to reproduce copyright material.

Acknowledgements

Every effort has been made to trace all copyright holders, but if any have been inadvertently overlooked, the Publishers will be pleased to make the necessary arrangements at the first opportunity.

Although every effort has been made to ensure that website addresses are correct at time of going to press, Hodder Education cannot be held responsible for the content of any website mentioned in this book. It is sometimes possible to find a relocated web page by typing in the address of the home page for a website in the URL window of your browser.

Hachette UK's policy is to use papers that are natural, renewable and recyclable products and made from wood grown in sustainable forests. The logging and manufacturing processes are expected to conform to the environmental regulations of the country of origin.

Orders: please contact Bookpoint Ltd, 130 Park Drive, Milton Park, Abingdon, Oxon OX14 4SE. Telephone: (44) 01235 827720. Fax: (44) 01235 400454. Email education@bookpoint.co.uk Lines are open from 9 a.m. to 5 p.m., Monday to Saturday, with a 24-hour message answering service. You can also order through our website: www.hoddereducation.co.uk

ISBN: 978 1 5104 1767 0

First published in 2018 by
Hodder Education,
An Hachette UK Company
Carmelite House
50 Victoria Embankment
London EC4Y 0DZ

www.hoddereducation.co.uk

Impression number   10 9 8 7 6 5 4 3 2 1

Year   2022 2021 2020 2019 2018

Cover photo © Katerina Kovaleva/123RF.COM

Typeset in Bembo Std Regular 11/13 by Integra Software Services Pvt. Ltd., Pondicherry, India

Printed in Spain by Graphycems

A catalogue record for this title is available from the British Library.

Get the most from this book

Welcome to your Revision Guide for the Applied Mathematics content of the MEI AS Level in Mathematics course (OCR Mathematics B). This book will provide you with reminders of the knowledge and skills you will be expected to demonstrate in the exam with opportunities to check and practice those skills on exam-style questions. Additional hints and notes throughout help you to avoid common errors and provide a better understanding of what's needed in the exam. In order to revise the Pure Mathematics content of the course, you will need to refer to My Revision Notes: OCR B (MEI) A Level Mathematics Year 1/AS (Pure).

The material in this book is also relevant for students sitting the MEI A Level Mathematics exam. However, students may prefer to use the Revision Guide for the Applied Mathematics content of the MEI A Level Mathematics course (OCR Mathematics B) which covers all the applied mathematics needed for the exam.

Included with the purchase of this book is valuable online material that provides full worked solutions to all the 'Target your revision', 'Exam-style questions' and 'Review questions', as well as full explanations and feedback to each answer option in the 'Test yourself' multiple choice questions. The **online material** is available at **www.hoddereducation.co.uk/ myrevisionnotesdownloads**.

Features to help you succeed

Target your revision

Use these questions at the start of the two sections (one on Statistics, the other on Mechanics) to focus your revision on the topics you find tricky. **Short answers** are at the back of the book, but use the **worked solutions online** to check each step in your solution.

About this topic

At the start of each chapter, this provides a concise overview of its content.

Before you start, remember

A summary of the key things you need to know **before** you start the chapter.

Key facts

Check you understand all the key facts in each subsection. These provide a useful checklist if you get stuck on a question.

Worked examples

Full worked examples show you what the examiner expects to see in order to ensure full marks in the exam. The examples cover a sample of the type of questions you can expect.

Hint

Expert tips are given throughout the book to help you do well in the exam.

Common mistakes

Your attention is drawn to typical mistakes students make, so you can avoid them.

Test yourself

Succinct sets of multiple-choice questions test your understanding of each topic. Check your **answers online**. The **online feedback** will explain any mistakes you made as well as common misconceptions, allowing you to try again.

Exam-style questions

For each topic, these provide typical questions you should expect to meet in the exam. **Short answers** are at the back of the book, and you can check your working using the **online worked solutions**.

Review questions

After you have completed each of the two sections in the book, answer these questions for more practice. **Short answers** are at the back of the book, but the **worked solutions online** allow you to check every line in your solution.

At the end of the book, you will find some useful information:

Exam preparation

Includes hints and tips on revising for the AS Mathematics exam, and details the exact structure of the exam papers.

Make sure you know these formulae for your exam

Provides a succinct list of all the formulae you need to remember and the formulae that will be given to you in the exam.

Please note that the formula sheet as provided by the exam board for the exam may be subject to change.

During your exam

Includes key words to watch out for, common mistakes to avoid and tips if you get stuck on a question.

My revision planner

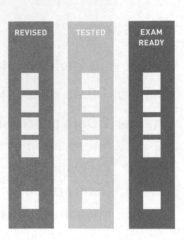

Review questions (Mechanics)

Answers to Target your revision, Exam-style questions and Review questions

Go online for:
- full worked solutions and answers to the Test yourself questions
- full worked solutions to all Exam-style questions
- full worked solutions to all Review questions
- full worked solutions to the Target your revision questions

www.hoddereducation.co.uk/myrevisionnotesdownloads

STATISTICS

Target your revision (Statistics)

1 Understand the terms population and sample

Insert the words 'population' and 'sample' in the correct places in the following sentences. Each word may occur more than once. Statistical methods use data from a .... to infer results for the .... Larger............s tend to be more representative of the ....

(see page 6)

2 Use simple random sampling

Describe how to take a simple random sample of 40 students from a secondary school which has 900 students.

(see page 6)

3 Understand sampling methods: opportunity sampling, systematic sampling, stratified sampling, quota sampling, cluster sampling, self-selected samples

A student wants to know why visitors come to a town. She asks all the people getting off a particular long distance train arriving at the town. Which sampling method is this?

(see pages 6–7)

4 Recognise possible sources of bias when sampling

An advice centre is only open part time. The leaders want to know whether the opening times are suitable for the people who want to use the centre. They ask all the people who come to the centre one week whether the opening times are suitable. Identify the bias in this method of collecting data.

(see pages 8–9)

5 Be aware of practicalities of implementation of sampling methods

A marine biologist suspects that the cod in the North Sea tend to be smaller than they used to be. She wants to take a sample of the cod and measure them. Explain why it would be impossible to use simple random sampling.

(see pages 8–9)

6 Recognise categorical, discrete, continuous and ranked data

Which of the words: categorical, discrete, continuous, ranked would apply to students' responses to a question about what they ate for breakfast that day?

(see pages 13–14)

7 Interpret a bar chart

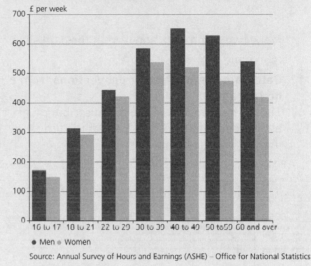

Median full-time gross weekly earnings by sex and age group, UK, April 2015

● Men ● Women

Source: Annual Survey of Hours and Earnings (ASHE) – Office for National Statistics

Lily says, 'The graph shows that women's earnings decrease after the age of 39.' Is Lily correct? Justify your answer.

(see pages 15–16)

8 Interpret a dot plot

The dot plot shows the number of people living in each of a random sample of houses in a town.

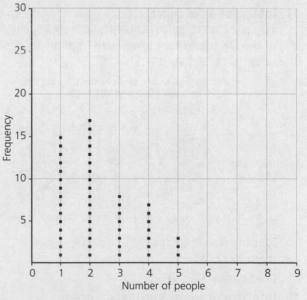

There were 50 houses in the sample. Estimate the percentage of houses in the town with two or fewer people living in them.

(see page 14)

9 Interpret a histogram or frequency chart

The following histogram shows the ages at which men died in France in 1944.

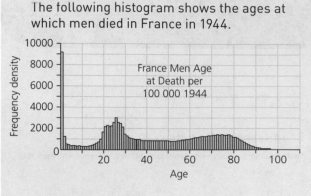

France Men Age at Death per 100 000 1944

Make two comments about what the data show.

(see page 26)

10 Calculate frequencies or proportions from a histogram

The histogram shows the age distribution of all residents of Cheshire East at the time of the 2011 census.

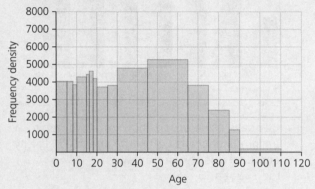

Estimate the total number of people living in Cheshire East in 2011.

(see page 26)

11 Interpret a pie chart

The pie charts below show the percentage of the UK workforce employed in different sectors in 1992 and 2011.

Percentage share of workforce by sector, April to June 1992 and 2011, UK

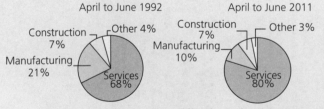

April to June 1992

Construction 7%
Other 4%
Manufacturing 21%
Services 68%

April to June 2011

Construction 7%
Other 3%
Manufacturing 10%
Services 80%

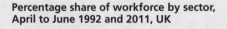

Source: Labour Force Survey – Office for National Statistics

Identify two changes that have taken place from 1992 to 2011.

(see pages 13–14)

12 Interpret a stem-and-leaf diagram

Two groups of people completed the same puzzle. The stem-and-leaf diagram below shows the time each person took to complete the puzzle.

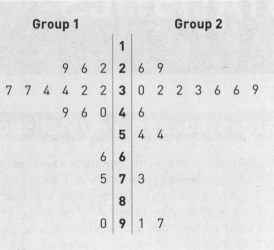

Group 1		Group 2
	1	
9 6 2	2	6 9
7 7 4 4 2 2	3	0 2 2 3 6 6 9
9 6 0	4	6
	5	4 4
6	6	
5	7	3
	8	
0	9	1 7

Key

8 | 2 means 82 seconds for group 2

Compare the times taken by the two groups.

(see page 16)

13 Interpret box plots

The box plots below show the prices paid for houses in two towns in England in June 2016.

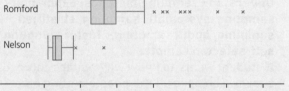

Romford

Nelson

Compare house prices in Nelson and Romford. Make three distinct comments.

(see page 37)

14 Interpret a cumulative frequency diagram

The cumulative percentage diagram below shows the age distribution of the population of Japan in 2016.

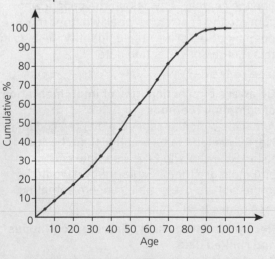

Estimate the median age in Japan in 2016.

(see page 40)

15 Recognise symmetrical, unimodal, bimodal and skewed distributions

The histogram below shows the percentage age distribution of Japan.

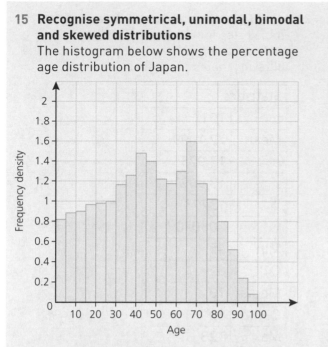

Describe the age distribution.

(see pages 15–16)

16 Interpret a scatter diagram and best fit model, describe correlation

The scatter diagram below shows the physician density and life expectancy at birth for all the countries of the world. A line of best fit has been drawn using a spreadsheet.

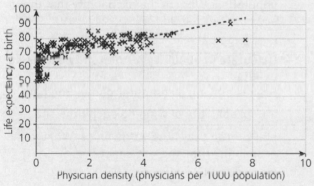

i Describe the correlation in the scatter diagram.

ii Is the line of best fit a good model for the relationship between physician density and life expectancy at birth? Give a reason for your answer.

(see page 49)

17 Recognise outliers and different groups in scatter diagrams

The scatter diagram below shows the diastolic and systolic blood pressure in mm Hg of a sample of American adults.

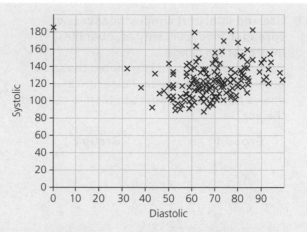

i Identify any outliers in the scatter diagram.

II For each outlier, determine whether it is likely to be the result of an error.

(see page 50)

18 Select or critique data presentation techniques

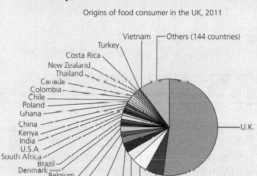

Origins of food consumer in the UK, 2011

Based on the farm-gate value of unprocessed food

i What is wrong with the way the data have been presented above?

ii Give one way to improve the presentation.

(see page 14)

19 Calculate and interpret measures of central tendency: median, mode, mean, mid-range

A business publishes data on annual earnings of its staff as shown in the table below.

Salary band (£)	Male frequency	Female frequency
21 000–25 000	9	9
27 000–38 000	7	5
40 000–55 000	10	2
60 000–100 000	1	0

i   Estimate the male mean salary.

ii Estimate the female mean salary

iii Comment on your answers.

(see page 23)

20 Use and interpret simple measures of spread: range, percentiles, quartiles, interquartile range

Two groups of people completed the same puzzle. The stem and leaf diagram below shows the time each person took to complete the puzzle.

Group 1		Group 2
	1	
9 6 2	2	6 9
7 7 4 4 2 2 2	3	0 2 2 3 6 6 9
9 6 0	4	6
	5	4 4
6	6	
5	7	3
	8	
0	9	1 7

Key

8 | 2   means 82 seconds for group 2

i   For group 1, find the median and interquartile range.

ii For group 2, find the median and interquartile range.

iii Compare the times for groups 1 and 2 by using your calculated values of median and interquartile range.

(see page 36)

21 Calculate and interpret variance and standard deviation

A business publishes data on annual earnings of its staff as shown in the table below.

Salary band (£)	Male frequency	Female frequency
21 000–25 000	9	9
27 000–38 000	7	5
40 000–55 000	10	2
60 000–100 000	1	0

i   Estimate the standard deviation of male salary.

ii Estimate the standard deviation of female salary

iii Comment on your answers.

(see page 46)

22 Identify outliers and clean data

Pulse rates for a sample of adults are analysed using software which produces the following table of statistics.

Statistics	
n	191
Mean	72.1099
σ	13.9255
s	13.9621
Σx	13773
Σx^2	1030209
Min	45
Q1	62
Median	70
Q3	80
Max	128

Determine whether any of the pulse rates could be considered to be outliers.

(see pages 46–47)

23 Calculate probability for independent events

12% of the population are left handed. Two people sit next to each other on a train. Assume they are chosen at random from the population. What is the probability that they are both left handed?

(see page 62)

24 Calculate probability for mutually exclusive events

Fran is playing a computer game. She has a $\frac{1}{3}$ probability of winning each time she plays. Fran plays the game twice. What is the probability that she wins once?

(see page 61)

25 Recognise situations which give rise to a binomial distribution

A class of 20 students includes 10 who come to college by bus. 5 students will be picked from the class at random to take part in a survey. Will the binomial distribution with $n=5$, $p=0.5$ be a suitable model for the number of those picked who travel by bus? Justify your answer.

(see page 74)

26 Calculate probabilities for a binomial distribution

$X \sim B(34, 0.3)$. Calculate $P(15 \leqslant X \leqslant 20)$.

(see page 76)

27 Calculate expected value and expected frequencies for a binomial distribution

A survey estimates that 5% of the population are vegetarians. Assume that this percentage is correct. What is the expected number of vegetarians in a random sample of 2000 drawn from the population?

(see pages 59, 76)

28 Calculate probabilities for a simple discrete random variable

Probabilities for a discrete random variable are shown in the table below in terms of a constant, k.

x	0	1	2
$P(X = x)$	$3k$	$5k$	$2k$

Find the value of k.

(see page 66)

29 Understand terminology associated with hypothesis testing

The significance level of a hypothesis test is 5%. Rhys says this means that there is a 5% chance that the test will give the wrong result; this is not quite right. What does a significance level of 5% mean?

(see page 82)

30 Know when to apply 1-tail or 2-tail tests when setting up a binomial hypothesis test

In one area of the country, 55% voted to leave the European Union in the June 2016 referendum. A journalist suspects that views in that area have changed so she asks a random sample of 100 voters how they would vote today. 52 say they would vote to leave. She conducts a hypothesis test with $H_0: p = 0.55$ where p is the proportion who would vote to leave the European Union. What should the alternative hypothesis be?

(see pages 83 and 89)

31 Identify suitable null and alternative hypotheses for a binomial hypothesis test

A telephone helpline claims that 90% of callers wait less than 5 minutes to speak to an adviser. An inspector suspects that fewer than 90% are spoken to within 5 minutes and carries out a hypothesis test using a random sample of callers. Write down suitable null and alternative hypotheses for the test.

(see page 83)

32 Find a critical region for a binomial hypothesis test

The peppered moth has a dark variety and a light variety. In one area, statistics from five years ago show that there are 20% light moths in the area. A biologist thinks this may have changed. She will examine a random sample of 50 moths to conduct a hypothesis test at the 5% level with the following hypotheses.
$H_0: p = 0.2$
$H_1: p \neq 0.2$ where p is the proportion of peppered moths which are light.
Find the critical region for the test.

(see page 91)

33 Conduct a hypothesis test using the binomial distribution and interpret the results in context

The owners of a plate factory want to know whether changes in production have decreased the proportion of faulty plates. The proportion of faulty plates has been 8%. A statistician conducts a hypothesis test at the 5% significance level with the following hypotheses.
$H_0: p = 0.08$

$H_1: p < 0.08$ where p is the proportion of faulty plates.
The p-value for the test is 1.55%. What should the statistician conclude?

(see pages 84–85)

Short answers on page 157

Full worked solutions online

CHECKED ANSWERS

Chapter 1 Data collection

About this topic

Statistical methods are widely used in other subjects and in the workplace to gain insight into situations. Increasing use of computers has made data easier to collect, store and share; this has increased the potential uses of statistics. Working with real data will involve deciding which of the statistical techniques you know are appropriate.

Before you start, remember ...

● Knowing what the mode, median and mean of a data set are.
● Interpreting bar charts and pie charts.

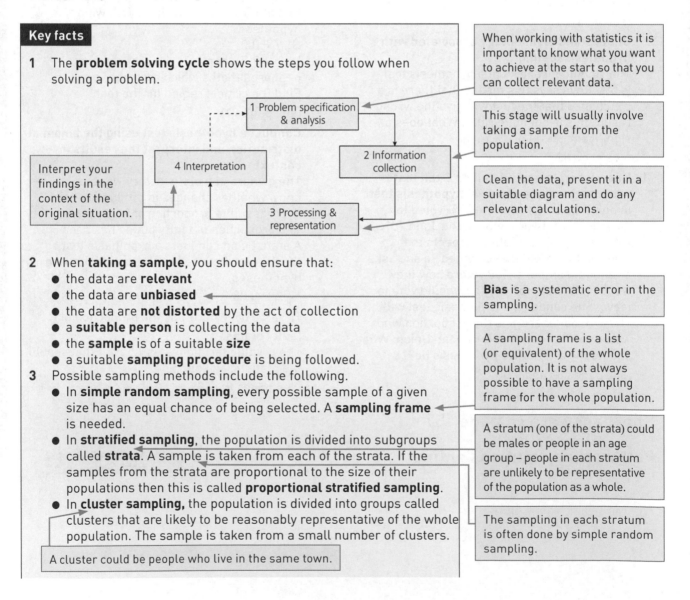

Key facts

1 The **problem solving cycle** shows the steps you follow when solving a problem.

> When working with statistics it is important to know what you want to achieve at the start so that you can collect relevant data.

> This stage will usually involve taking a sample from the population.

> Clean the data, present it in a suitable diagram and do any relevant calculations.

- 1 Problem specification & analysis
- 2 Information collection
- 3 Processing & representation
- 4 Interpretation

Interpret your findings in the context of the original situation.

2 When **taking a sample**, you should ensure that:
- the data are **relevant**
- the data are **unbiased**
- the data are **not distorted** by the act of collection
- a **suitable person** is collecting the data
- the **sample** is of a suitable **size**
- a suitable **sampling procedure** is being followed.

> **Bias** is a systematic error in the sampling.

3 Possible sampling methods include the following.
- In **simple random sampling**, every possible sample of a given size has an equal chance of being selected. A **sampling frame** is needed.
- In **stratified sampling**, the population is divided into subgroups called **strata**. A sample is taken from each of the strata. If the samples from the strata are proportional to the size of their populations then this is called **proportional stratified sampling**.
- In **cluster sampling,** the population is divided into groups called clusters that are likely to be reasonably representative of the whole population. The sample is taken from a small number of clusters.

> A sampling frame is a list (or equivalent) of the whole population. It is not always possible to have a sampling frame for the whole population.

> A stratum (one of the strata) could be males or people in an age group – people in each stratum are unlikely to be representative of the population as a whole.

> The sampling in each stratum is often done by simple random sampling.

> A cluster could be people who live in the same town.

- In **systematic sampling**, the **sampling frame** is ordered and then individuals are chosen at regular intervals.

The first individual in a systematic sample is often chosen at random.	The choice of who is sampled to fulfil the quota is usually left to the interviewer doing the survey.	The sampling frame could be a list of the population on a spreadsheet.

- In **quota sampling**, the number to be sampled from each **stratum** of the population is decided. This method is often used for surveys.
- **Opportunity sampling** is used when a sample is readily available.
- A **self-selected sample** consists of volunteers.

> For example, all the people at a meeting.

4 **Cleaning data** includes recognising and dealing with errors, outliers and missing data.
An **outlier** is a data item that is not consistent with the rest of the data. There are some rules for checking whether a data item is an outlier but sometimes it is obvious.

Worked examples

Example 1

A researcher wants to investigate the future career plans of A level Mathematics students.
The table below lists six things which may be relevant to the investigation.

Questionnaire	Measurement	A level Mathematics students in England
A sample of A level Mathematics students in Birmingham	A list of all A level Mathematics students in England	A level Mathematics students at a revision day

From the table, identify each of the following.

i The population.

ii The sampling frame.

iii An opportunity sample.

iv A cluster sample.

v A possible method of collecting relevant data.

> The researcher would need to decide whether he/she wanted to investigate students in a particular year or over time.

Solution

i The population is A level Mathematics students in England.

ii The sampling frame is a list of all A level Mathematics students in England.

iii A level Mathematics students at a revision day is an opportunity sample.

> **Common mistake:** The students in this sample can all be easily surveyed at once – the students in Birmingham could not, so the revision day is the opportunity sample.

iv A sample of A level Mathematics students in Birmingham is a cluster sample.

v A questionnaire is a possible method of collecting relevant data.

> **Common mistake:** Measurement can be a way of collecting relevant data but not in this case – there is nothing to measure.

Example 2

The dot plot below shows the number of mm of rainfall collected by a rain gauge in Leeds on each day in July 2013.

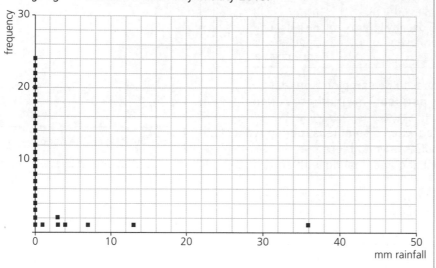

i   State the mode.

ii Find the median.

iii Identify an outlier.

iv Is the outlier necessarily a mistake in the data? Explain your reasoning.

v   The data were collected as part of an investigation into whether July in Leeds is getting wetter than it used to be. What other data would be needed?

Solution

i   The mode is zero.

ii The median is zero.

iii The value which is about 35 mm is an outlier.

iv No, the outlier need not be a mistake. It is possible to have very rainy days in July.

v   Rainfall data for Leeds for July for other years.

> Most of the days had no rain (over 20 days) so this is the mode

> There are 31 days in July. If the amounts of rainfall were put in order, the 16th number would be the median. There are over 20 days with no rain so the 16th number must be zero.

> The value which is about 13 mm could also be considered an outlier

> Outliers can be mistakes in the data but they need not be. Sometimes a value might be so high that it is clearly not possible but that is not the case here.

> When choosing a sample, you need to know what population you are selecting from. In this case, it is rainfall in Leeds in July over time.

Advantages and drawbacks of sampling methods

Method	Advantages	Drawbacks
Simple random sampling	• avoids bias • probability can be used to study properties of samples	• needs a sampling frame • may be expensive or time-consuming to collect the sample

Method	Advantages	Drawbacks
Stratified sampling	• ensures all subgroups in the population are represented	• may be expensive or time-consuming to collect the sample • it may be hard to find out information about which subgroups individuals are in
Cluster sampling	• does not need a sampling frame • reduces cost and time spent in sample collection	• it's hard to know whether the sample is representative of the whole population or whether it is biased
Systematic sampling	• easy and quick when a sampling frame is available	• needs a sampling frame • could be biased due to a pattern in the sampling frame
Quota sampling	• ensures all specified subgroups in the population are represented	• it can be biased depending on how the individuals are chosen
Opportunity sampling	• quick and easy	• the sample may not represent the population
Self-selected sampling	• quick and easy	• very likely to be biased

Worked example

Example 3

A journalist on the Avonford Star wants to know what the public thinks of a new government policy so he invites people to add comments to the Avonford Star website.

   i What type of sample is this?
   ii Give two reasons why the sample may be biased.

Hint: The people who respond decide to do so for themselves so they are choosing to be in the sample rather than being chosen by someone else. This makes it self-selected.

Solution

   i *This is a self-selected sample.*

   ii *Possible reasons include the following:*

     • *The sample excludes people who do not read the Avonford Star.*
     • *The sample excludes people who do not use the internet.*
     • *People may be influenced by seeing other people's comments.*
     • *Only people with strong views are likely to respond.*

Hint: Notice that there are four possible reasons given here. The question only asks for two reasons. Any two sensible reasons will do. You should not give more than two reasons. When asked to do something in an examination you need to follow instructions carefully.

Test yourself

1 A researcher investigating health has a sample of data about adult heights. The data give the height and sex of each person but individuals cannot be identified. The height of one man in the sample is listed as 6 m. What should she do with this data value?

   A She should include it – it would bias the sample to ignore it.

   B She should delete it from the data set – it is clearly wrong.

   C She should not use the sample – it contains errors.

   D She should go and measure the height of the man herself.

   E She should realise that it is meant to be 6 ft and use that value for the height.

2 A student is investigating how students in his college use their leisure time. He starts with a list of all students at the college and looks at male and female students separately. He chooses equal numbers of male students and female students at random. Which sampling method is the student using?

   A Cluster sampling

   B Quota sampling

   C Simple random sampling

   D Stratified sampling

   E Systematic sampling

3 A biologist is investigating how many eggs a particular species of bird lays. She visits three sites where she knows the birds nest and counts the number of eggs in a random sample of nests at each site. What sampling method is the biologist using?

   A Stratified sampling

   B Simple random sampling

   C Quota sampling

   D Opportunity sampling

   E Cluster sampling

4 Which of the following is NOT true of simple random sampling?

   A Simple random sampling avoids bias.

   B A sampling frame is needed for simple random sampling.

   C A simple random sample will always be more representative of the population than any other type of sample.

   D Probability methods can be used to study properties of random samples.

   E If individuals chosen as part of a simple random sample refuse to take part then this may introduce bias.

5 A market researcher wants to know what users of a particular post office think of the facilities available. Which of the following is likely to be the least biased sampling method?

   A Choose everyone who is in the post office at noon on Monday.

   B Put an advert in the local paper and ask volunteers to get in touch if they use the post office and want to give their opinions about the facilities.

   C Visit the post office at 12 times during one week, two on each day Monday to Saturday, and choose every fifth person who comes into the post office during half an hour.

   D Ask post office staff to give a letter to 20 men and 20 women who use the post office. The letter asks the person who receives it to complete a questionnaire and give it back to the post office staff.

   E Put up a sign that can be seen by everyone entering the post office asking them to get in touch with the market researcher to give their opinion about the facilities. Choose a simple random sample from those who get in touch.

Full worked solutions online

Exam-style question

A journalist wants to find out how sixth form students in Leeds intend to vote in the first general election after they are allowed to vote.

i   What is the population for this investigation?

ii Describe how to take a simple random sample of size 100 from the population.

Instead of taking a random sample of size 100, the journalist decides to visit a politics class in one of the sixth form colleges in Leeds and ask the students in the class to put up their hands for the political party they would vote for.

iii State the sampling method being used.

iv Identify two possible sources of bias with this method of collecting data.

Short answers on page 158

Full worked solutions online

CHECKED ANSWERS

Chapter 2 Data processing, presentation and interpretation

About this topic

Statistics is about making sense of data; knowing what kind of data you have is important so that you can use appropriate techniques to help you unlock the information they contain. This chapter deals with statistical graphs and summary statistics.

Being able to understand statistical graphs and use them appropriately is an important skill. You are likely to encounter such graphs in the news, in other subjects and also in your future work. Software, for example a spreadsheet, is often used to produce statistical graphs but care is needed to ensure that the correct type of graph is used to display the data and that the axes and labels are correct to ensure that a misleading impression is not given.

Averages are often used in everyday life so you need to understand them thoroughly. Averages are sometimes called "measures of central tendency". Averages are particularly useful for summarising and comparing larger sets of data; these are usually presented as frequency tables. In addition to knowing the average, it is also important to know how spread out the data are. The interquartile range and standard deviation are commonly used measures of spread.

Histograms are usually used for continuous data which have been grouped; understanding them will give you a useful way of displaying data. You may already have studied them at GCSE but do check your understanding as it is quite easy to go wrong with histograms.

Cumulative frequency curves are often used to estimate the median, quartiles and percentiles of grouped data; these measures are used to compare individuals or groups of people. For example, growth curves for young children use them. Box-and-whisker diagrams use the median and quartiles to provide a visual way of comparing sets of data.

Scatter diagrams show relationships between two variables; considering how two variables are related to each other is important in statistical modelling to make predictions.

Before you start, remember ...

- How to draw bar charts and pie charts from GCSE.
- Putting numbers in order (including decimals and negative numbers) from GCSE.
- Frequency tables, cumulative frequency and rounding from GCSE.
- Correlation from GCSE.
- Straight line graphs.

Interpreting graphs for one variable

Key facts

- Categorical or qualitative data are not numerical.
- Numerical or quantitative data can be subdivided into discrete or continuous.
- The data may also be the ranks of the variables in the set.
- Continuous data may take any values; these are usually within a range.
- Discrete data may only take certain separate values, for example whole numbers.
- Vertical line charts and dot plots are often used for discrete numerical data. The vertical axis shows the frequency. The horizontal axis is a continuous scale.
- Bar charts are often used for categorical data. There should be gaps between the bars.
- A compound bar chart stacks the bars for several sets of data on top of each other so they can be compared.
- A multiple bar chart puts the bars for several sets of data next to each other so they can be compared.
- Pie charts illustrate the frequencies of sections of the data set in comparison with each other and the whole set.
- You can use a stem-and-leaf diagram for discrete or continuous data.
- A stem-and-leaf diagram shows the shape of the distribution as well as keeping the original data values.
- The position of the 'tail' of the data shows you the kind of skewness.
- A distribution with one peak is unimodal. If it has two peaks, it is bimodal; the peaks need not be the same height.

Types of data

Categorical or **qualitative** data are in categories; they are not numerical. Examples include eye colour, breed of dog and favourite type of book.

Numerical or **quantitative** data consist of numbers.

Continuous variables are measured on a scale, e.g. length, weight, temperature. For any two possible values, you can always find another possible value between them. In practice, continuous data are always rounded because there is a limit to how accurately they can be measured.

Discrete variables are often just whole numbers; e.g. the number of children in a family but they need not be e.g. shoe sizes sometimes include half sizes.

> **Common mistake:** Sometimes categorical data are coded using numbers but the number is just a shorthand, it does not make the data numerical.

Using statistical diagrams

Statistical diagrams can give a quick visual impression of a whole set of data. It is important to use the right kind of diagram and to draw it correctly so that the impression given is an accurate one. The flow chart below will help you to understand when different graphs and charts are appropriate for discrete and continuous numerical data.

Start in the grey box towards the centre to help you make a decision about the best diagram to use. Sometimes there is more than one possibility.

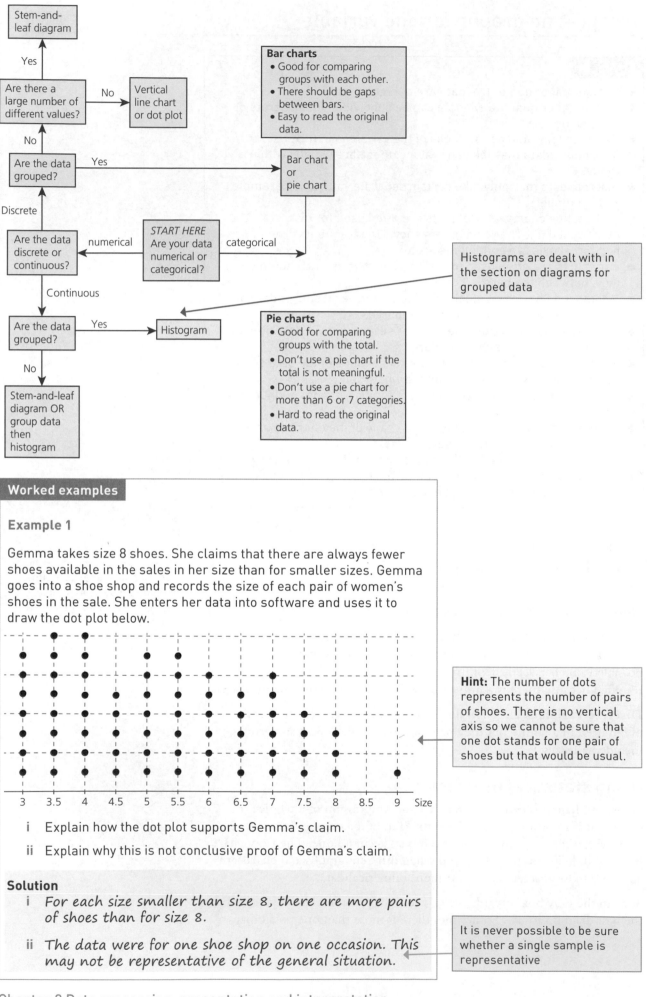

Stem-and-leaf diagram

Yes

Are there a large number of different values? → No → **Vertical line chart or dot plot**

No

Are the data grouped? → Yes → **Bar chart or pie chart**

Discrete

Are the data discrete or continuous? ← numerical ← **START HERE Are your data numerical or categorical?** → categorical → (to Bar chart or pie chart)

Continuous

Are the data grouped? → Yes → **Histogram**

No

Stem-and-leaf diagram OR group data then histogram

Bar charts
• Good for comparing groups with each other.
• There should be gaps between bars.
• Easy to read the original data.

Histograms are dealt with in the section on diagrams for grouped data

Pie charts
• Good for comparing groups with the total.
• Don't use a pie chart if the total is not meaningful.
• Don't use a pie chart for more than 6 or 7 categories.
• Hard to read the original data.

Worked examples

Example 1

Gemma takes size 8 shoes. She claims that there are always fewer shoes available in the sales in her size than for smaller sizes. Gemma goes into a shoe shop and records the size of each pair of women's shoes in the sale. She enters her data into software and uses it to draw the dot plot below.

Hint: The number of dots represents the number of pairs of shoes. There is no vertical axis so we cannot be sure that one dot stands for one pair of shoes but that would be usual.

i Explain how the dot plot supports Gemma's claim.

ii Explain why this is not conclusive proof of Gemma's claim.

Solution

i *For each size smaller than size 8, there are more pairs of shoes than for size 8.*

ii *The data were for one shoe shop on one occasion. This may not be representative of the general situation.*

It is never possible to be sure whether a single sample is representative

Compound bar charts and multiple bar charts

Compound bar charts and multiple bar charts are used to compare two, or more, sets of similar data.

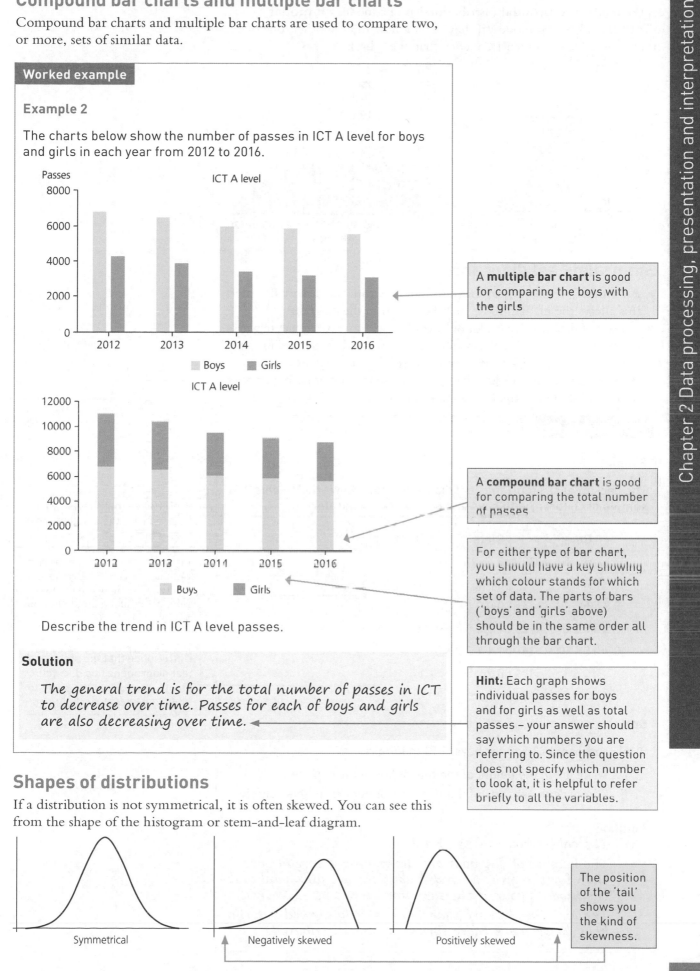

Worked example

Example 2

The charts below show the number of passes in ICT A level for boys and girls in each year from 2012 to 2016.

A **multiple bar chart** is good for comparing the boys with the girls

A **compound bar chart** is good for comparing the total number of passes

For either type of bar chart, you should have a key showing which colour stands for which set of data. The parts of bars ('boys' and 'girls' above) should be in the same order all through the bar chart.

Describe the trend in ICT A level passes.

Solution

> The general trend is for the total number of passes in ICT to decrease over time. Passes for each of boys and girls are also decreasing over time.

Hint: Each graph shows individual passes for boys and for girls as well as total passes – your answer should say which numbers you are referring to. Since the question does not specify which number to look at, it is helpful to refer briefly to all the variables.

Shapes of distributions

If a distribution is not symmetrical, it is often skewed. You can see this from the shape of the histogram or stem-and-leaf diagram.

Symmetrical   Negatively skewed   Positively skewed

The position of the 'tail' shows you the kind of skewness.

The value of the variable at the peak of a diagram showing frequency is the mode. A **unimodal** distribution has one mode; a **bimodal** distribution has two modes (if they are not next to each other, one of them could have a higher frequency than the other).

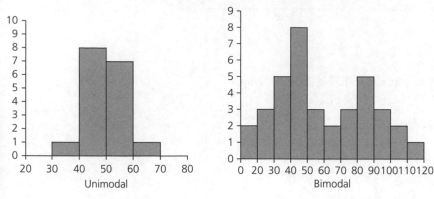

Unimodal   Bimodal

Stem-and-leaf diagrams

- An ordered stem-and-leaf diagram is drawn with the 'leaves' in order; this allows the median to be found easily from the diagram.
- For sets of data that have a lot of leaves on each row, two separate rows can be used for each 'stem' one for leaves 0 to 4 and the second for 5 to 9. Sometimes an asterisk is used on the first of the two stems.
- Back to back stem-and-leaf diagrams are useful for comparing two sets of data, as shown in the following example.

Worked example

Example 3

The back to back stem-and-leaf diagram below shows the heights of a sample of children, with boys and girls shown separately.

Boys		Girls
7 4 0	12	3 4
8 6 5 4	13	6 6 7
8 8 5 5 5 1	14	0 0 0 6
9 7 7 3 3 3 2 0	15	1 2 3 4 5 6 7 8 9
8 7 5 3 1 0	16	1 1 1 2 2 4 5 9
6 1	17	0 2 2 5
8 0	18	

Hint: Notice that the leaves are ordered with the smallest near the stem

To help you see the skew, you can turn the diagram on its side, with the smallest stem on the left

Hint: Notice that the stem-and-leaf diagram has no decimal points. You look at the scale to find out where they should go

Scale

0|15|1 represents 1.5 m for boys; 1.51 m for girls

i   Write down the height of the tallest boy in the sample.

ii   Compare the shapes of the two distributions from this sample.

Solution
i   *The tallest boy is 1.88 m tall.*
ii   *The heights of the boys are slightly more spread out than for the girls. Both distributions are unimodal and the modal group is the same for each (150–159 cm).*

The girls' heights are slightly negatively skewed but the boys' heights are more symmetrically distributed.

Test yourself

1 A head of year wants to illustrate how many days absence each of six classes had in a school year. Each class has the same number of students. He needs to decide whether to draw a vertical line chart, a bar chart or a pie chart. Which of these kinds of statistical diagram would be well suited for this purpose?

Form	Days absence
10A	217
10B	74
10F	99

A   Only a vertical line chart.

B   Only a bar chart.

C   Only a pie chart.

D   Either a vertical line chart or a bar chart but not a pie chart.

E   Either a bar chart or a pie chart but not a vertical line chart.

2 The nicotine content in mg for each of a sample of cigarettes is shown in the stem-and-leaf diagram below.

```
 4 | 3            Scale
 5 | 7            6| 8 means 0.68 mg
 6 | 8
 7 | 4 8
 8 | 0 6 7
 9 | 0 5 6
10 | 2 4 7 8
11 | 2 7
12 | 6
```

What is the shape of the distribution?

A   positively skewed

B   negatively skewed

C   symmetrical

D   bimodal

E   none of these.

3 The pie chart illustrates the number of bedrooms in houses for sale advertised in the Avonford Star. There were 72 houses for sale altogether. How many had one bedroom?

A   1.8

B   2

C   10

D   50

E   Not possible to tell.

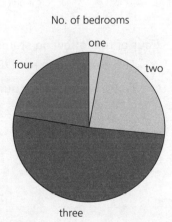

No. of bedrooms

4 Four of the following statements about the graph below are false and one is true. Which one is true?

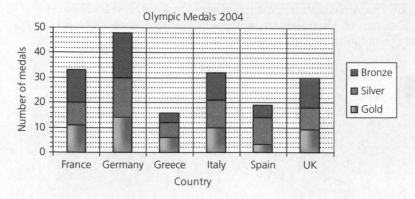

A UK had 30 bronze medals.

B Greece had more silver medals than gold medals.

C France had more silver medals than the UK.

D Germany had more silver medals than gold medals.

E Ranking these countries in order of their number of gold medals gives the same result as ranking them by their total number of medals.

5 The cloudiness of the sky is measured in oktas; the number of oktas is the fraction of the sky (in eighths) covered by cloud. The vertical line charts below show the frequencies of the different possible values for the cloud cover for the days from May to October in 1987 (left) and 2015 (right) for Heathrow.

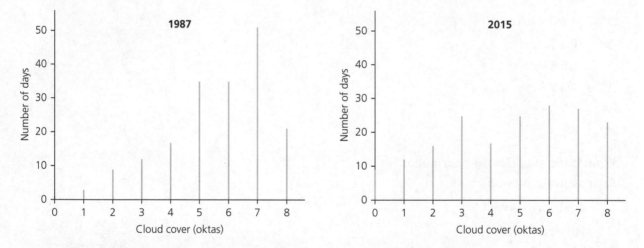

Four of the following statements are true and one is false. Which one is false?

A The distribution for 1987 is positively skewed.

B There were similar numbers of days in May to October in each of 1987 and 2015 with completely cloudy sky.

C   In 1987, over one quarter of the days for which data are shown had 7 okta cloud cover.

D   These data show that in both 1987 and 2015, more than half the sky was cloud covered on more than half of the days.

E   May to October 2015 was generally less cloudy than the same period in 1987.

Full worked solutions online

CHECKED ANSWERS

Exam-style question

The chart below compares the life expectancy at birth for males and females in the London borough of Westminster.

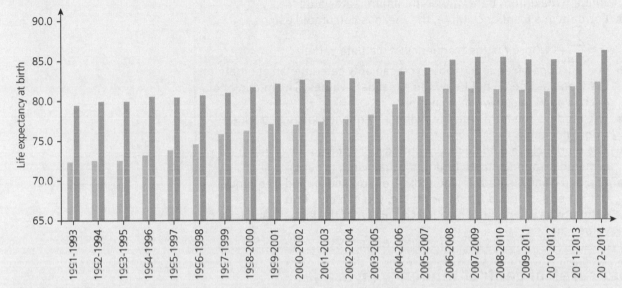

I   Give two ways in which the chart should be improved to make it a better representation of the data.

ii   Make two statements comparing the life expectancy at birth for males and females in Westminster.

iii Life expectancy at birth is calculated by using data about the ages at which people die. For small areas, data from several years are grouped. Explain why this gives better estimates of life expectancy than just looking at one year at a time.

iv   For 1992 to 1994, show that 76 is a reasonable estimate of life expectancy at birth for all people.

Short answers on page 158

Full worked solutions online

CHECKED ANSWERS

> **Key facts**
>
> - The mean (symbol $\bar{x}$) is found by adding the data values and dividing by the number of them; $\bar{x} = \dfrac{\sum x}{n}$.
> - The median is the middle value when the data are put in order.
> - When data values have been put in order, they are referred to as ranked data.
> - The mode is the most common item of data (modal class is used for grouped data).
> - The mid-range is half way between the lowest and highest data items.
> - The range is a measure of spread and not an average. The range = maximum data value – minimum data value.
> - For data in a frequency table, the mean is calculated using
> $\bar{x} = \dfrac{\sum xf}{n}$ where f is the frequency of the data value x.
> - For a grouped frequency table, you can only calculate an estimate of the mean as you do not know the exact data values. The midpoint of each group is used when doing this.
> - The mode is the item with the highest frequency in an ungrouped frequency table.
> - The modal class is the group with the highest frequency in a grouped frequency table (if it has equal width groups).
> - The mid-range is easy to calculate from an ungrouped frequency table. For a grouped frequency table, you can only find an estimate as you do not know for certain what the highest and lowest values are.

Measures of central tendency (averages)

A measure of central tendency is a single value that is used to represent a set of data. They are sometimes called averages and are useful for comparing sets of data; e.g. teachers might compare classes of students by looking at the average test mark for each class.

Mean

The mean is often referred to as 'the average' in everyday speech. It is worked out by adding up the data values and dividing by the number of them. This can be written as the formula:

$$\bar{x} = \frac{\sum\limits_{i=1}^{n} x_i}{n}$$

where

- $\bar{x}$ stands for the mean.
- The data items are $x_1, x_2 \ldots x_n$; x_i is a typical, or general, data item.
- The Greek letter sigma (Σ) means 'the sum of'. It gives the total of all the data.
- The number of data items is n.

The formula is often written more simply as:

$$\bar{x} = \frac{\sum x}{n}$$

Median

The median is the middle value when the data items are put in order. If there are an even number of data values, there will be two middle values; the median is half way between them.

Worked examples

Example 1

The part-time weekly earnings of nine students are shown below.

  i   Find the mean.

  ii   Find the median.

£10.28 £30.00 £49.20 £29.50 £0.00 £35.10 £58.50 £39.00 £20.00

> **Hint:** You are expected to use the statistical functions of your calculator to get the mean and median – make sure you know how to do this. You should round your final answer sensibly. Most calculators will give a list of statistics once you have input the data.

Solution

> Remember units

  i   The mean is £30.18 (nearest penny).

  ii   The median is £30.

The mean is a 'fair shares' average; the value you calculated in part **i** is the same as you would get working out how much each person would receive if they all shared their earnings equally among themselves.

You will usually find the mean by using statistical functions on a calculator but you need to know how it is worked out so that you can use the mean to calculate the total.

Example 2

There are 16 students in a class; 15 of them measure their pulse rates and find the mean is 65. The 16th student measures his pulse rate the following lesson; it is 120. What is the mean pulse rate for all 16 students?

Solution

The total for the 15 students is $15 \times 65 = 975$

The total for all 16 students is $975 + 120 = 1095$.

The mean for all 16 is $1095 \div 16 = 68.4375 = 68.44$ (2 d.p.)

> $\text{mean} = \dfrac{\text{total}}{n}$
>
> so
>
> $\text{total} = n \times \text{mean}$

> Even if all the data are whole numbers, the mean need not be

---

Mode

The mode is the most commonly occurring value. There can be more than one mode, or there might not be a mode. The mode can be used for categorical data; the other measures of central tendency cannot.

Mid-range

The mid-range is half way between the largest and smallest data values. It is very easy to work out but can be very unrepresentative if there is an unusually large (or small) data value. However, it is usually reasonably representative for symmetrical distributions and has the advantage of allowing you to make a very quick estimate of the average.

Worked example

Example 3

Find the mid-range of the earnings of the nine students in Example 1.

Solution

$$\text{Mid-range} = \frac{0 + 58.50}{2} = £29.25$$

Hint: To find the number half way between two numbers, add them and divide the answer by 2.

Common mistake: The mid-range is a measure of central tendency. Be careful not to confuse it with the range, which is a measure of spread.

Hint: All the measures of central tendency produce a value that represents the whole set of data. If you get an answer for an average of numerical data that is either larger than the largest value in the set or smaller than the smallest value, you know you have made a mistake somewhere.

Deciding which average to use

- The mean and the mid-range are affected by one very large (or small) data value.
- The mean allows you to work out the total of the data and uses all the values.
- The median is not affected by extreme data values.
- The mid-range is easy to calculate and gives the value midway between the highest and lowest values; it is useful for symmetrical distributions.
- The mode gives the most likely value to occur; it can be used for categorical data.
- For symmetrical, unimodal data the mean, median, mode and mid-range are all equal.

Averages from an ungrouped frequency table

Worked example

Example 4

The table below shows the number of goals scored by the winning team (or the goals scored by either team in the case of a draw) for a sample of league football matches.

Number of goals	0	1	2	3	4	5	6	7	8
Frequency	4	21	22	15	8	3	0	0	1

i   Find the mean.
ii Find the median.

If a data value has frequency zero, this means it did not occur. There were no games with 6 or 7 goals.

Common mistake: In a frequency table for numerical data, there are two sets of numbers. The frequency tells you how often each value occurred; the other numbers are the data values that you are dealing with. Understanding this will help you avoid some of the common mistakes made when working with frequency tables.

Solution

	List 1	List 2	List 3	Lis
SUB				
1	0	4		
2	1	21		
3	2	22		
4	3	15		

Hint: Your calculator should allow you to type in the data as a frequency table. Make sure you know how to do this for your calculator.

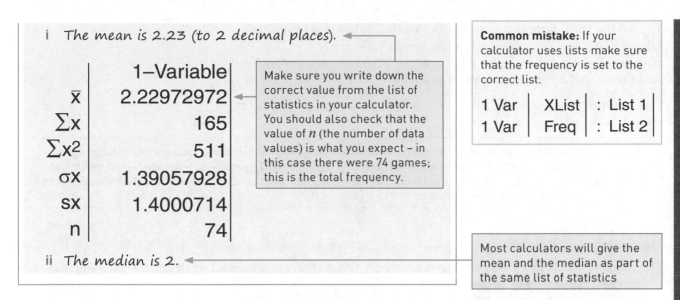

i *The mean is 2.23 (to 2 decimal places).*

1–Variable	
x̄	2.22972972
Σx	165
Σx²	511
σx	1.39057928
sx	1.4000714
n	74

Make sure you write down the correct value from the list of statistics in your calculator. You should also check that the value of n (the number of data values) is what you expect – in this case there were 74 games; this is the total frequency.

Common mistake: If your calculator uses lists make sure that the frequency is set to the correct list.

1 Var	XList	: List 1
1 Var	Freq	: List 2

ii *The median is 2.*

Most calculators will give the mean and the median as part of the same list of statistics

For the data in Example 4, it is easy to see that the **mode** is two goals, because this happened 22 times (the highest frequency).

It is also easy to work out the **mid-range** because you can see that the highest value is 8 and the lowest value is 0. So the mid-range is 4.

Averages from a grouped frequency table

For the **mean**, you do not know all the individual data values so you use the midpoint for each group as an estimate of all the values in it; this means that your eventual answer is an estimate of the mean.

The table below shows the time of the first goal for a sample of league football matches, together with the midpoint of each group.

Using the midpoint of the group as an estimate for all the data values in the group loses information; this does not matter too much for the estimation of the mean as that is based on the total of all data values.

Time (minutes)	Frequency, f	Midpoint, x
1–10	9	5.5
11–20	12	15.5
21–30	10	25.5
31–45	19	38.0
46–60	11	53.0
61–70	6	65.5
71–80	3	75.5
81–90	0	85.5

Hint: The mean can be found using a calculator in the same way as for an ungrouped table **except** that you are working with the midpoint of each group as an estimate of the values in that group. It is a good idea to write down the midpoints to help you type them into the calculator correctly.

The mean is 32.8 (to 3 s.f.).

Hint: It can be easy to go wrong when working out mean from a frequency table; always look back at the data to see whether your final answer is a reasonable average.

Common mistake: It would not make sense to state the **modal class** for this table because there are some groups wider than others; if they were the same width, things could look different.

Common mistake: The times have been rounded before being put into the frequency table. The 11–20 minute group is really 10½ ≤ time < 20½ minutes; in this case the midpoint would be the same so it does not make a difference. Be especially careful when working with ages; they are always rounded down. 11–20 years would really be 11 ≤ age < 21 years, with a midpoint of 16.0.

To find the **mid-range**, you need to know the highest and lowest data values. You know the lowest data value for the table above is in the 1–10 minutes group and the highest value is in the 71–80 minutes group. You could use 1 and 80 to estimate the mid-range.

The **median** is estimated by using a cumulative frequency graph; this is revised in the next two main sections.

> **Common mistake:** When you enter the midpoints and frequencies into your calculator, you are using the midpoint of each group as an estimate of all data values in that group; this does **NOT** give a good estimate for the median of grouped data. The same applies to the quartiles. Only the mean, standard deviation and various totals should be used from the list of statistics which the calculator gives for grouped data.

Test yourself

TESTED

1 A sample of students are asked how many school dinners they ate in the past week with the following results.

Number of dinners eaten	0	1	2	3	4	5
Frequency	7	10	10	12	21	18

Four of the statements below are false and one is true. Which one is true?

A The mid-range is 2.5.

B The mode is 10.

C The number of students in the sample was 6.

D The number of students in the sample was 15.

E The distribution shows positive skewness.

2 A group of students were asked how many books they have in their room and gave answers as follows:

5   5   3   2   1   3   2   57   2   9   3   8

Four of the following statements are false and one is true. Which one is true?

A The mode is 2.5.

B The mean should not be used for these data because it is impossible to have 8.3333... books.

C There is no limit to the number of books a student could have so the data are continuous.

D The median, which is 3, is a good average to use as it is unaffected by the large value of 57.

E The mid-range is $56 \div 2 = 28$.

3 A class of students consists of 9 girls and 20 boys. The mean weight of the girls is 53 kg; the mean weight of the boys is 61 kg. What is the mean weight for the whole class?

A   3.9 kg   B   55.5 kg   C   57 kg   D   58.5 kg   E   848.5 kg

4 A teacher has to choose someone from a class to represent them in a regional spelling competition. They have regular spelling tests (marked out of 10) in the class and two students, who have taken all 20 of the class tests, are willing to represent the class. Information about their average scores in the class tests is shown below. Four of the following statements are true and one is false. Which one is false?

	Mark	Lucy
Mean	6.8	7.6
Mode	8	6

A Because there were 20 class tests, each of them must have scored their mode more than twice.

B Their means show that Mark scores less than 7 half the time and Lucy scores more than 7 half the time.

C The distribution of Mark's scores cannot be positively skewed or symmetrical.

D For a short competition, Mark is a better choice than Lucy as his higher mode shows that he is more likely to do well.

E For a long competition, Lucy is a better choice than Mark as her higher mean shows that her total score is higher than his.

5 The waist measurements of a sample of boys are shown below.

Waist (cm)	50–59	60–69	70–79	80–89	90–109
Frequency	2	45	80	19	7

Four of the statements below are false and one is true. Which one is true?

A   A good estimate of the mid-range is 30 cm.

B   A good estimate of the mid-range is 70 cm.

C   A good estimate of the mid-range is 84 cm.

D   The median is 74.5 cm.

E   The median is somewhere in the group 70–79 cm.

Full worked solutions online

CHECKED ANSWERS

Exam-style question

The table below shows the ages of students on a course at a university.

Ages	17–21	22–26	27–31	32–40
Frequency	13	11	3	4

i   Alan is calculating an estimate of their mean age. He starts by saying that the midpoint of the first group is 19. Is this correct? Explain your answer.

ii Find an estimate of the mean age.

iii Why is your answer only an estimate?

iv One of the students who was put in the last group in the table should not have been put in this group as he is 55 years old. Find a better estimate of the mean taking account of this information.

Short answers on page 158

Full worked solutions online

CHECKED ANSWERS

Diagrams for grouped data

REVISED

Key facts

1 Histograms
- Histograms are usually used for continuous grouped data.
- The horizontal axis on a histogram is an even scale showing the values of the variable (e.g. height, wage, age).
- The vertical axis shows frequency density; this is calculated by dividing the frequency by the class width.
  - The label can be 'frequency density' or, for example, 'frequency per cm'
  - The symbol / can be used instead of 'per'.
- For a histogram with equal width bars, software often puts frequency on the vertical axis rather than frequency density; such a diagram is a **frequency chart**.
- Because the horizontal axis on a histogram is a continuous scale, there are no gaps between bars.
- The frequency represented by a bar in a histogram is given by the area of the bar.
- The modal class is the group with the highest frequency density.

2 Cumulative frequency graphs
- Points on a cumulative frequency graph are plotted at the upper boundary of the group.
- Cumulative frequency tells you how many data items were up to a particular value.

Histograms for continuous data with unequal class widths

Worked example

Example 1

The birth weights of a sample of babies are given in the table below. Draw a histogram to illustrate these data.

Weight (kg)	Frequency	Class width (kg)	Frequency density
$2.0 \leqslant w < 2.5$	2	0.5	4
$2.5 \leqslant w < 3.0$	13	0.5	26
$3.0 \leqslant w < 3.25$	15	0.25	60
$3.25 \leqslant w < 3.5$	14	0.25	56
$3.5 \leqslant w < 3.75$	10	0.25	40
$3.75 \leqslant w < 4.0$	10	0.25	40
$4.0 \leqslant w < 4.5$	11	0.5	22

Using narrower groups where there is more data (in the middle for this example) can help to show an appropriate level of detail

Hints: Frequency density is calculated by dividing the frequency by the class width; it is useful to have a column in the table for the class widths.

Solution

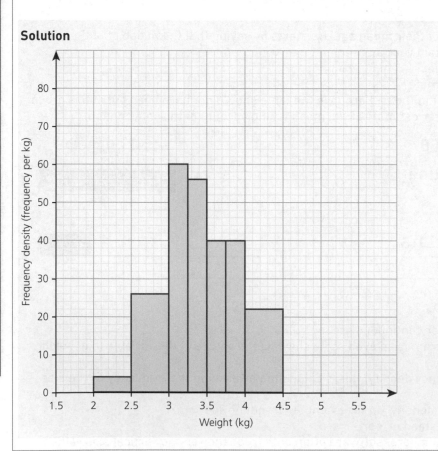

Hints:
- The area of a bar in a histogram is the frequency of the group it represents.
- The vertical axis could be labelled either 'frequency density' or 'frequency per kg'.

Common mistake: When drawing a histogram for **age**, remember that, for example a group aged 14–19 years could have people that are just a day short of their 20th birthday so its width would be 20 – 14 = 6 years.

Cumulative frequency graphs

The birth weights of a sample of babies are given in the table below, together with the cumulative frequencies. This is the same data as for Example 1.

Weight (kg)	Frequency	Cumulative frequency
$2.0 \leqslant w < 2.5$	2	2
$2.5 \leqslant w < 3.0$	13	15
$3.0 \leqslant w < 3.25$	15	30
$3.25 \leqslant w < 3.5$	14	44
$3.5 \leqslant w < 3.75$	10	54
$3.75 \leqslant w < 4.0$	10	64
$4.0 \leqslant w < 4.5$	11	75

The entries in the cumulative frequency column are found by adding up the frequencies as you go down the table. In this example the cumulative frequency tells you how many babies were in the given weight group and the ones below it.

Hint: The final cumulative frequency gives the total frequency.

The cumulative frequency graph is shown below.

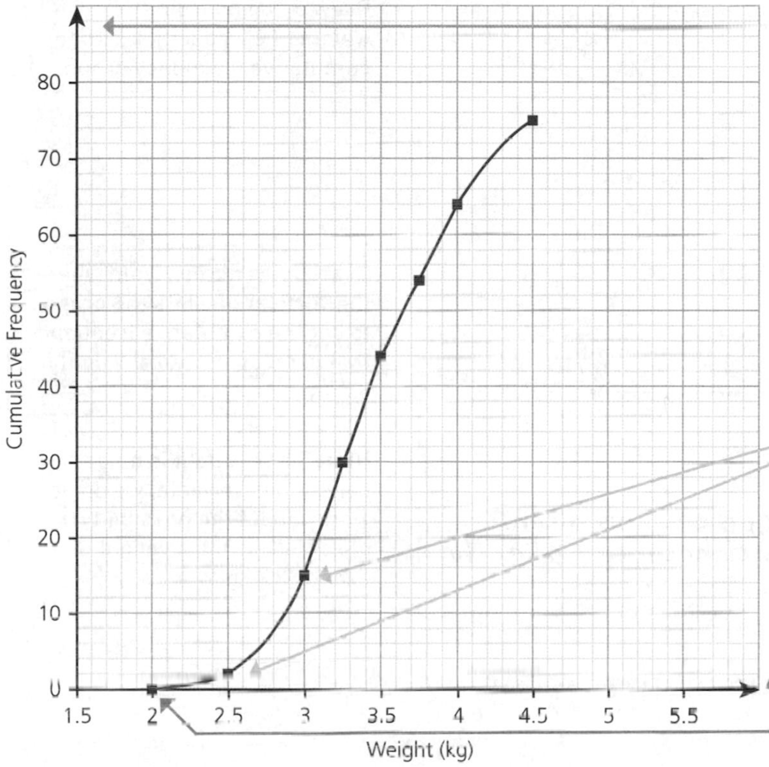

Weight (kg)

The vertical axis shows cumulative frequency.

Common mistake: The first cumulative frequency in the table is 2; this means that two babies had a weight up to 2.5 kg. The second cumulative frequency is 15; this means that 15 babies had a weight up to 3 kg. The cumulative frequency should always be plotted at the upper limit of the group.

The horizontal axis is an even scale showing the variable being studied.

Hints: There were no babies with weight below 2 kg so the cumulative frequency for 2 kg is 0.

Worked example

Example 2

Use the cumulative frequency graph to estimate the number of babies in the sample with a birth weight of over 3.4 kg.

Solution

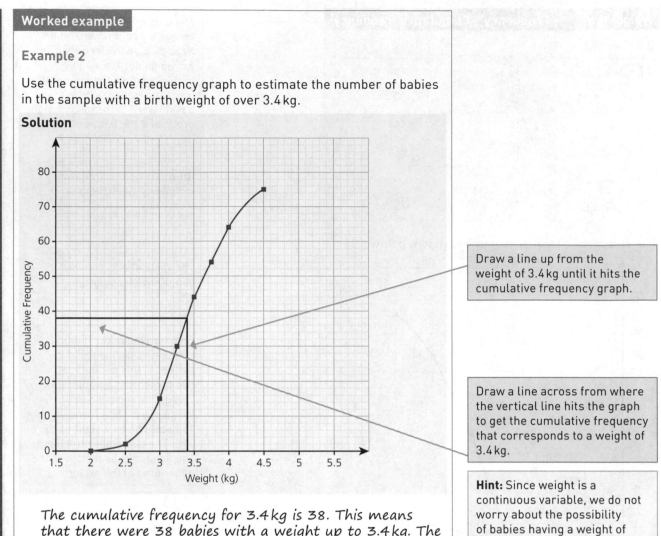

Draw a line up from the weight of 3.4 kg until it hits the cumulative frequency graph.

Draw a line across from where the vertical line hits the graph to get the cumulative frequency that corresponds to a weight of 3.4 kg.

Hint: Since weight is a continuous variable, we do not worry about the possibility of babies having a weight of exactly 3.4 kg.

The cumulative frequency for 3.4 kg is 38. This means that there were 38 babies with a weight up to 3.4 kg. The number with weights over 3.4 kg is 75 – 38 = 37.

The relationship between the cumulative frequency graph and the histogram

The histogram and the cumulative frequency graph for the data in Example 1 are shown side by side below.

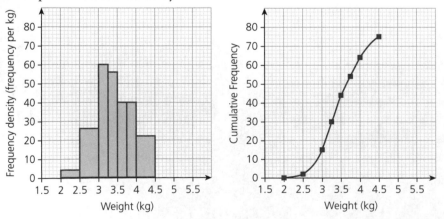

Hint: For high bars, the cumulative frequency goes up more quickly as you go across the bar so the cumulative frequency graph for that range is steeper.

The area in the histogram represents frequency. The area to the left of a vertical line in the histogram is the cumulative frequency and so is the same as the height of the cumulative frequency curve for that value of the variable.

Test yourself

1 A sample of students was asked to estimate the length of a line. The data are shown in the table.
Four attempts at a histogram for these data are shown below. Three of them are incorrect and one is correct. Find the one which is correct.

Estimate (cm)	$14 \leqslant l < 18$	$18 \leqslant l < 22$	$22 \leqslant l < 24$	$24 \leqslant l < 26$	$26 \leqslant l < 30$	$30 \leqslant l < 34$	$34 \leqslant l < 42$
Frequency	12	7	13	7	8	7	6

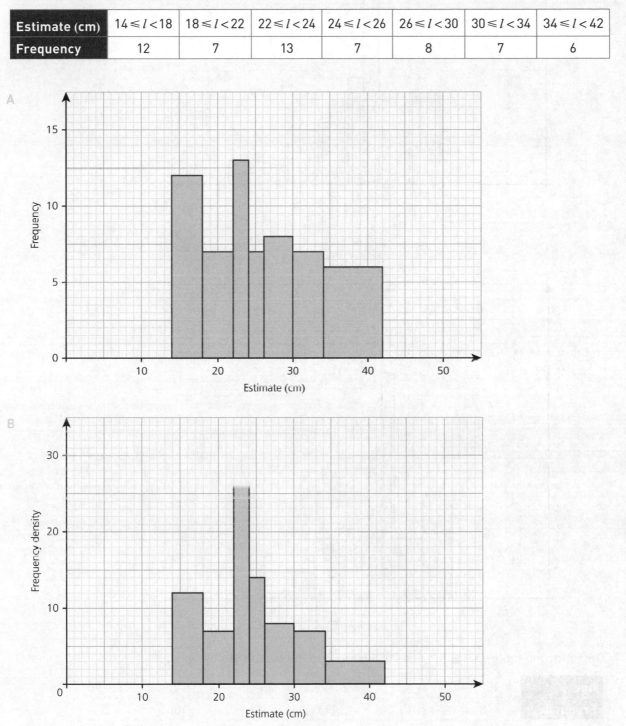

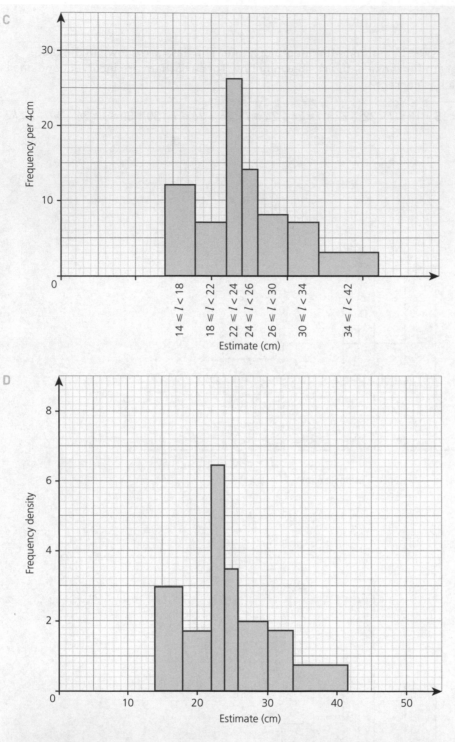

2 The following table shows the ages of a sample of the customers of a shop.

Age	13–16	17–19	20–25	26–35	36–55	56–70
Frequency	8	15	19	17	15	5

Between what limits should the next to last bar in the histogram be drawn?

A   35 to 55

B   36 to 56

C   35.5 to 55.5

D   35 to 56

E   36 to 55

3 The histogram below shows the pulse rates of a sample of people. How many people were in the sample?

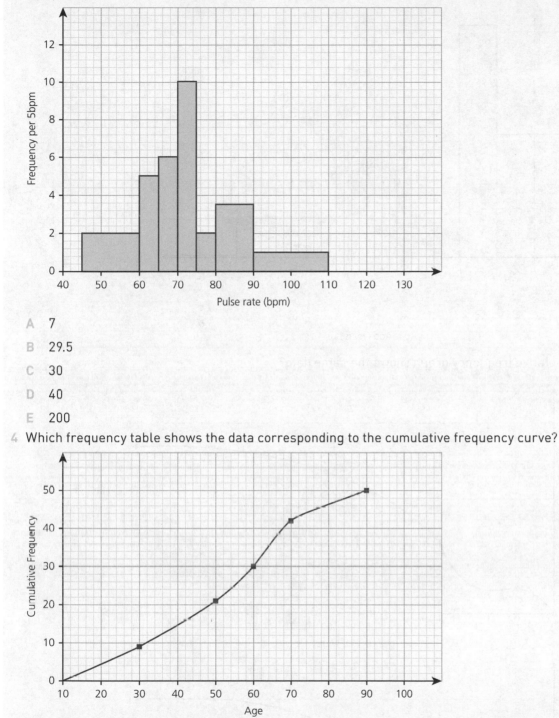

A 7

B 29.5

C 30

D 40

E 200

4 Which frequency table shows the data corresponding to the cumulative frequency curve?

	A		B		C		D
Age	Frequency	Age	Frequency	Age	Frequency	Age	Frequency
10–30	9	10–29	9	20–40	9	11–30	9
31–50	21	30–49	12	40–55	12	31–50	21
51–60	30	50–59	9	55–65	9	51–60	30
61–70	42	60–69	12	65–80	12	61–70	42
71–90	50	70–89	8	80–100	8	71–90	50

5 The histogram below shows the age distribution of the population of a city.

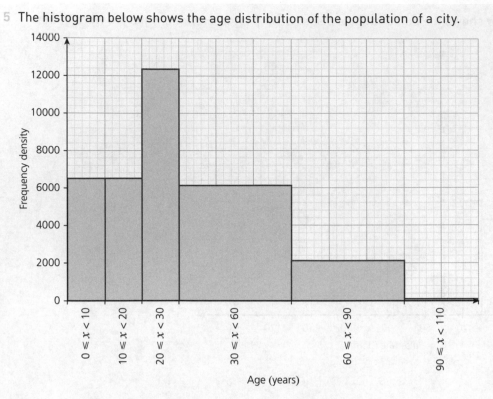

Which cumulative frequency graph shows the same data?

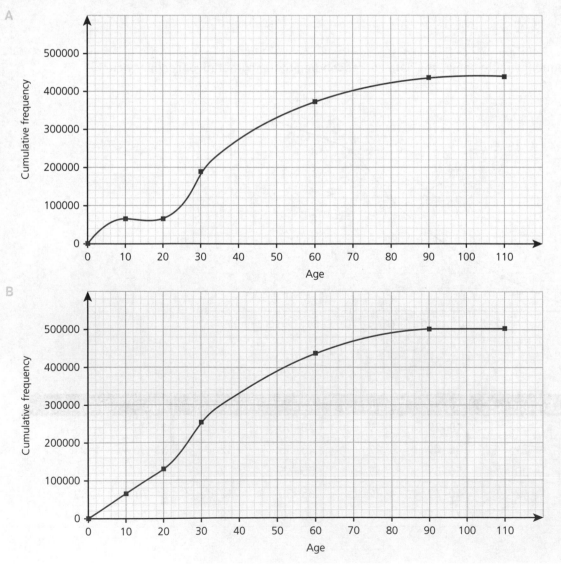

C

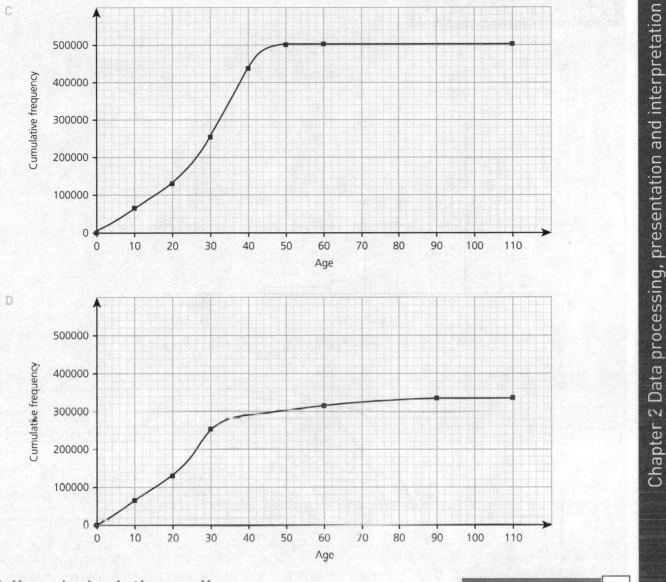

D

Full worked solutions online

CHECKED ANSWERS

Exam-style question

The histogram shows the pulse rates for male adults who took part in a health survey.

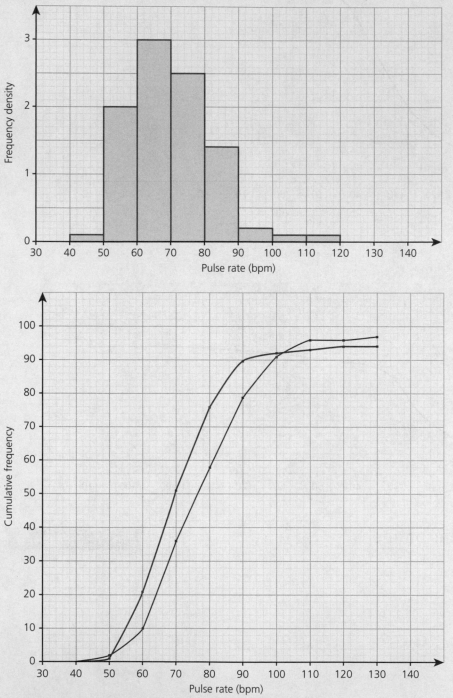

i   One of the cumulative frequency graphs is for the same pulse rate data as the histogram. Determine which one.

The other cumulative frequency graph is for the pulse rates for the female adults in the same survey.

ii   Compare the total numbers of males and females in the survey.
iii a Find the number of males with pulse rates between 90 and 100 bpm.
     b Find the number of females with pulse rates between 90 and 100 bpm.
iv   Make a general comment comparing male and female pulse rates.

Short answers on page 158

Full worked solutions online

CHECKED ANSWERS

Median and quartiles

Key facts

- The median is the middle value of a set of data.
- In an ordered data set of size n the median is the value of the $\frac{n+1}{2}$ th data item. When n is odd this is an actual data value; for n even, the median will be the mean of two values.
- You can find which group the median is in for a grouped frequency table in the same way.
- To find the quartiles for a small data set, find the median first. Find the middle value of the values below the median for the lower quartile. Find the middle value of the values above the median for the upper quartile.
- The median divides a histogram into two halves of equal area.
- The lower quartile cuts the histogram with one quarter of the area to the left of it; the upper quartile cuts the histogram with three quarters of the area to the left of it.
- To estimate the median from a cumulative frequency graph, read off the value with a cumulative frequency of $\frac{1}{2}n$, where n is the total number of data values.
- The upper and lower quartiles are estimated by reading off the values with cumulative frequencies of $\frac{3}{4}n$ and $\frac{1}{4}n$, respectively.
- In a box-and-whisker diagram, the box shows the distance between the quartiles, the whiskers show the rest.
- The interquartile range (IQR) is a measure of spread:
  IQR = upper quartile – lower quartile.
- A data value may be considered to be an outlier if it is more than $1.5 \times IQR$ above the upper quartile, or more than $1.5 \times IQR$ below the lower quartile.

Median and quartiles for a list of values

Worked example

Example 1

Find the median and quartiles of this data set: 6 4 6 7 8 11 2

Solution

You can find the median and quartiles by putting the data set into your calculator but it will be helpful if you understand how these values are calculated.

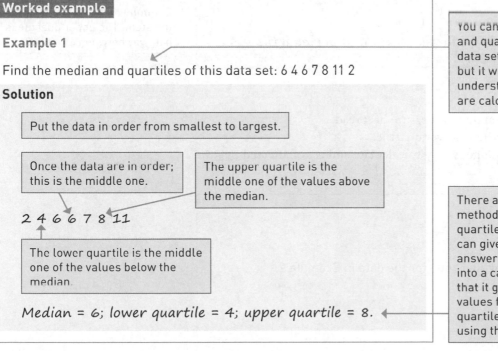

Put the data in order from smallest to largest.

Once the data are in order; this is the middle one.

The upper quartile is the middle one of the values above the median.

2 4 6 6 7 8 11

The lower quartile is the middle one of the values below the median.

Median = 6; lower quartile = 4; upper quartile = 8.

There are different recognised methods for deciding where quartiles are located; these can give slightly different answers. When you enter data into a calculator, you may find that it gives slightly different values for the lower and upper quartiles than you would get using the method above.

With an even number of data items, the median will be between two values.

1 3 3 4 5 5 6 8

Hint: The data are already in order; the middle is between 4 and 5. The median will be half way between 4 and 5 so 4.5.

Hint: The lower quartile is the middle one of the four values below the median. So it is midway between the two 3s and so it is 3.

Hint: The upper quartile is the middle one of the four values above the median. Again, it lies between two numbers so will be half way between 5 and 6.

Median = 4.5; lower quartile = 3; upper quartile = 5.5.

Hint: To find the number half way between two numbers, add them and divide the answer by 2.

Median and quartiles from a stem-and-leaf diagram

The data values in an ordered stem-and-leaf diagram are in order.

Worked example

Example 2

The stem-and-leaf diagram below shows the ages of 21 office staff. Find the median and quartiles.

Solution

```
1 | 8
2 | 1 5 6 6|7          Key
3 | 0 1 2 2 3 6 7 7    2|1 means 21 years old
4 | 1 1|3 4 7
5 | 2
6 | 7
```

This is the median; there are ten data items below it and ten above it.

The positions of the lower and upper quartiles are shown by vertical lines.

Median = 33; lower quartile = 26.5; upper quartile = 42.

Common mistake: Remember the stem. The upper quartile is half way between 41 and 43.

Interquartile range (*IQR*)

○ The interquartile range is a measure of spread.
  IQR = upper quartile − lower quartile.
○ The interquartile range is easy to calculate and is not affected by extreme values.

Worked example

Example 3

Calculate the interquartile range for the data in Example 2.

Solution

IQR = 42 − 26.5 = 15.5

Outliers

A data item may be considered an outlier if it is more than $1.5 \times IQR$ above the upper quartile or less than $1.5 \times IQR$ below the lower quartile.

Common mistake: There is another way of seeing whether a data item is an outlier using mean and standard deviation. Use the one that is easiest to work with in any situation but don't mix them up.

Worked example

Example 4

Could any of the ages in Example 2 be regarded as outliers?

Solution

$1.5 \times IQR = 1.5 \times 15.5 = 23.25$

$LQ - 23.25 = 26.5 - 23.25 = 3.25;$
$UQ + 23.25 = 42 + 23.25 = 65.25.$

Ages below 3.25 or above 65.25 can be considered to be outliers so 67 is an outlier.

You know what the quartiles are so it is easiest to use the method for outliers that uses quartiles.

Box-and-whisker diagrams

The data from Example 2 are shown in this box-and-whisker diagram.

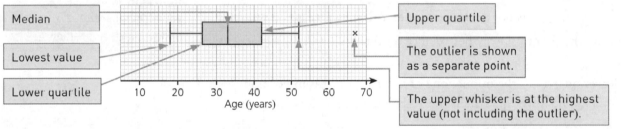

Median

Lowest value

Lower quartile

Upper quartile

The outlier is shown as a separate point.

The upper whisker is at the highest value (not including the outlier).

Skewness in box-and-whisker diagrams

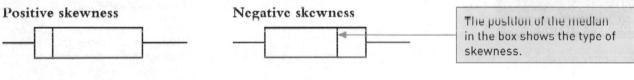

Positive skewness

Negative skewness

The position of the median in the box shows the type of skewness.

Median and quartiles from a table of data

The table below shows the number of goals scored by the winning team (or the goals scored by either team in the case of a draw) for a sample of league football matches.

Number of goals	0	1	2	3	4	5	6	7	8
Frequency	4	21	22	15	8	3	0	0	1

The median for this data set was found using a calculator in Example 4 of on page 22. The working for median and quartiles without a calculator is shown here to help with understanding working with grouped data.

To find the **median** from a table of data, start by adding a column for cumulative frequency to the table.

Number of goals	Frequency	Cumulative frequency
0	4	4
1	21	25
2	22	47
3	15	62
4	8	70
5	3	73
6	0	73
7	0	73
8	1	74

> The data are the number of goals and these are in order in the table.

> The cumulative frequency is 'the total frequency so far'.

> 74 is the total frequency for this table.

There are 74 data values in total. The median is in position number $\frac{74+1}{2} = 37.5$; this means it is half way between the 37th and 38th data values. The cumulative frequency column shows you that both the 37th and 38th values are two goals; so the median is two goals.

> **Common mistake:** There are 74 data values; half of 74 is 37. Don't make the mistake of thinking that the 37th data value is the median. There will be 37 data values below the median and 37 above the median so the median is between two data values.

To find the median of a data set of size n, start by asking whether n is odd or even.

- If n is odd, the **median** is in position number $\frac{n+1}{2}$ once the data values are in order.
- If n is even, $\frac{n+1}{2}$ is not a whole number, this means that you will be finding the point midway between two values.

> **Hint:** For n individual data values numbered 1, 2, ..., n the middle one is at position $\frac{n+1}{2}$.

To find the lower quartile, there are 37 data values below the median. The lower quartile is the middle one of these 37, i.e. it is in position $\frac{37+1}{2} = 19$. The lower quartile is the 19th data value so it is one goal.

By symmetry, the upper quartile is the 19th data value after the median. The first data value after the median is number 38.

> Notice that the data value is 37 more than the number after the median. So the upper quartile is data value $19 + 37 = 56$.

Data value	38	39	...	56
Number after median	1	2	...	19

The upper quartile is 3.

Estimating the median from a histogram

> **Hint:** For a grouped frequency table, you can find which groups the median and quartiles lie in using the same method as above.

Worked examples

Example 5

The birth weights of a sample of babies are given in the histogram and table below. Find an estimate of the median.

Weight (kg)	Frequency
2–2.5	2
2.5–3	13
3–3.25	15
3.25–3.5	14
3.5–3.75	10
3.75–4	10
4–4.5	11
TOTAL	75

> This is the same data as for Example 1 of on page 26.

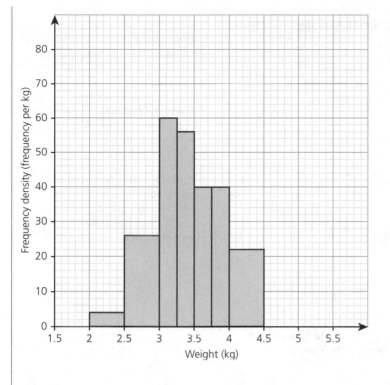

Solution

A vertical line at the median would cut the area of the histogram into two equal halves. To find an estimate of the median, start by finding the area of each bar. The frequency is the area so you can use the frequency as a measure of the area.

Half the total frequency is $75 \div 2 = 37.5$; this represents the area in front of the median. To get this area you need all of the first three bars and part of the fourth one.

$2 + 13 + 15 = 30$ ◄——— This is the total area of the first three bars.

$37.5 - 30 = 7.5$ ◄——— This is the area of the part of the fourth bar that is in front of the median.

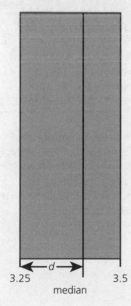

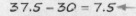

3.25   3.5
median

Hint: Don't worry about the units when finding the areas of the bars; they will look after themselves if your scales are correct. Make sure you find the area in the same way for each bar. You can count squares or work out width times height instead of using the frequency.

The fraction of the fourth bar that is needed is $\frac{7.5}{14}$ since its total frequency is 14.

The width of the bar is 0.25 so

$$\frac{7.5}{14} = \frac{d}{0.25}$$

$$d = \frac{0.25 \times 7.5}{14} = 0.13 \ (2\,d.p.)$$

median $= 3.25 + d = 3.25 + 0.13$

median $= 3.38\,kg$

Hint: The fraction of width is the same as the fraction of area.

Common mistake: The median is a weight so put units on the final answer.

Hint: The data have been grouped so you don't know what all the original values were. The final answer will be an estimate of the median.

Using cumulative frequency curves

Worked example

Example 6

The cumulative frequency graph for the data of Example 5 is shown below. Find estimates of the median and interquartile range.

Solution

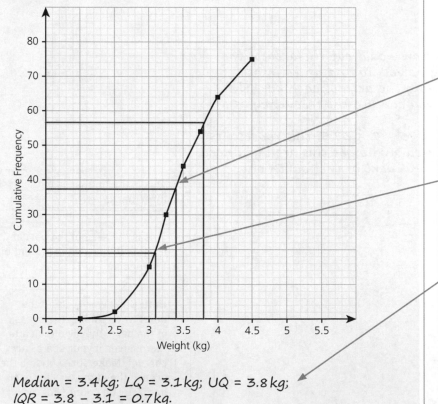

To estimate the median, read across from the cumulative frequency that is half the total: $(75 \div 2 = 37.5)$.

To find an estimate of the *LQ*, read across from the cumulative frequency that is one quarter of the total: $(75 \div 4 = 18.75)$.

These answers from reading from a graph are only to one decimal place, whereas the median obtained by calculating from the histogram was to two decimal places but they are only estimates as you do not know the actual data values.

Hint: For both the histogram and the cumulative frequency graph, the median is at the location of the middle of the continuous scale 0 to n and so it is estimated to be at $\frac{n}{2}$.

Median = 3.4 kg; LQ = 3.1 kg; UQ = 3.8 kg;
IQR = 3.8 − 3.1 = 0.7 kg.

Percentiles

The quartiles (including the median) divide the data into four equal groups. Percentiles divide the data into 100 equal groups.

Worked example

Example 7

Estimate the 90th percentile of the data in Example 6.

Solution

90% of 75 is 67.5.

> First find the relevant percentage of the cumulative frequency.

> **Common mistake:** 67.5 is not the 90th percentile. It tells you the cumulative frequency to look up.

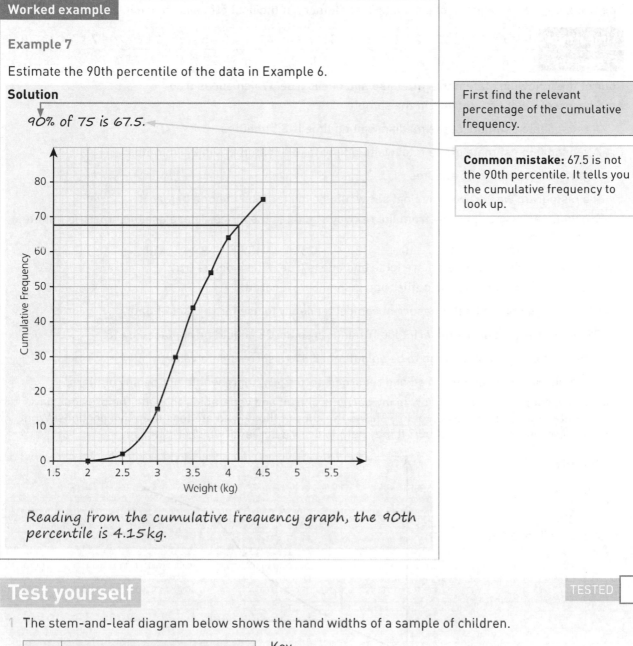

Reading from the cumulative frequency graph, the 90th percentile is 4.15 kg.

Test yourself

TESTED

1 The stem-and-leaf diagram below shows the hand widths of a sample of children.

8	6 6 6 7 7 7 0 0 0 0 0 0 9 9 9 9 9
9*	0 1 1 1 1 2 2 3 4 4 4
9	5 5 6 7 8 9
10*	2

Key

9|5 means

9.5 cm

Each stem has two rows of leaves; the first row has an asterisk (*) on the stem.
Which box-and-whisker diagram correctly shows the same data?

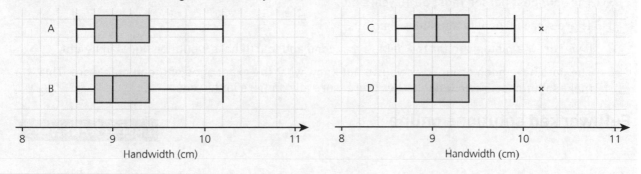

2 The waiting time, in minutes, of a sample of customers at the post office is shown in the table.

Time (t min)	$0 < t \leqslant 1$	$1 < t \leqslant 2$	$2 < t \leqslant 3$	$3 < t \leqslant 4$	$4 < t \leqslant 5$	$5 < t \leqslant 6$	$6 < t \leqslant 7$	$7 < t \leqslant 10$
Frequency	6	8	15	13	5	4	4	5

Four of the following statements are false and one is true. Which one is true?

A   There were eight customers in the sample.

B   A reasonable estimate of the median waiting time is 3.5 minutes.

C   A reasonable estimate of the total waiting time for all customers is 211 minutes.

D   The lower quartile is 15 minutes.

E   If a histogram was drawn, the final bar would be higher than the one before it.

3 Mr Brown times his bus journey from the railway station to the office. These are the times (in minutes) for ten journeys:
9   10   12   16   10   3   28   13   9   10
Four of the following statements are false and one is true. Which one is true?

A   The time of 28 minutes is an outlier and it should be ignored.

B   The mean is not the best measure of central tendency to use for this set of data.

C   28 minutes is probably a misprint for 18 minutes so the 28 should be replaced by 18.

D   3 minutes is not small enough to be an outlier so it is a correct data value.

E   If the times had been measured and recorded accurately, there would not be any outliers.

4 In a survey, people were asked how many portions of fruit and vegetables they ate the previous day. The cumulative frequency graph below shows the ages of those who ate less than five portions. What kind of skewness do the data show? It may help to draw a box-and-whisker diagram.

A   Positive.

B   Negative.

C   None.

D   Not possible to tell.

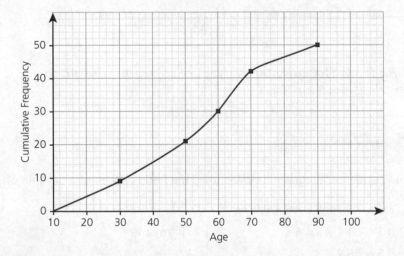

5 Here are five statements about the cumulative frequency curve in Question 4. Four of them are false and one is true. Find the one that is true.

A   The 90th percentile is 45.

B   The youngest person must be 10 years old.

C   There were 30 people aged less than 60.

D   If you drew a histogram for the data, the second and fourth bars would be equal in height.

E   The data have been grouped so you do not know what the exact ages of the people were. This makes it impossible to tell whether there were any outliers in this case.

Full worked solutions online

CHECKED ANSWERS

Exam-style question

As part of a biology experiment, a sample of leaves from a certain type of shrub is taken. The lengths of the leaves are measured. The results are summarised in the following table and cumulative frequency graph.

Length (x cm)	$2.0 \leqslant x < 3.0$	$3.0 \leqslant x < 4.0$	$4.0 \leqslant x < 4.5$	$4.5 \leqslant x < 5.0$	$5.0 \leqslant x < 6.0$	$6.0 \leqslant x < 8.0$
frequency	5	11	14	10	7	3

i   Find estimates of
  • the median,
  • the interquartile range.
ii Showing your working, suggest the lengths of leaves that might be considered outliers.
iii Alex only has the summary data in the table and graph. He says that there must be some leaves that were outliers in the original data. Explain why Alex is wrong.

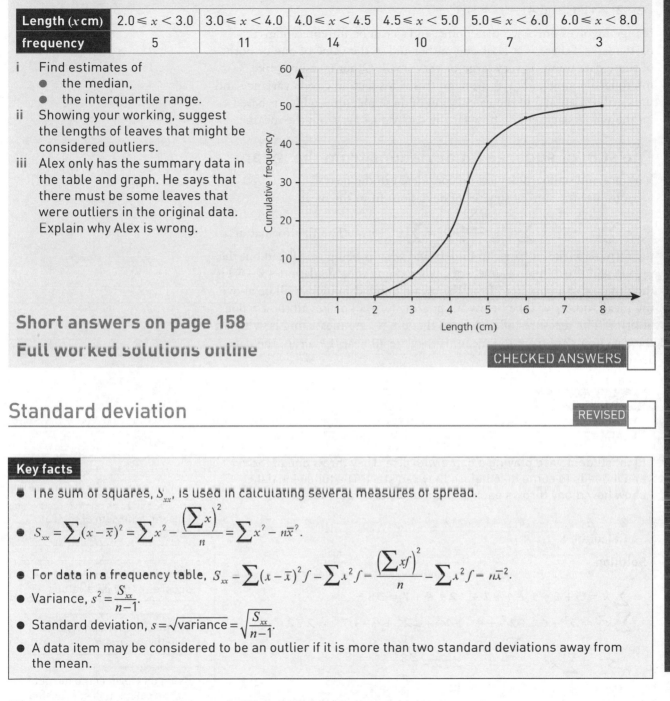

Short answers on page 158

Full worked solutions online

CHECKED ANSWERS

Standard deviation

REVISED

Key facts

• The sum of squares, S_{xx}, is used in calculating several measures of spread.

• $S_{xx} = \sum (x - \bar{x})^2 = \sum x^2 - \dfrac{\left(\sum x\right)^2}{n} = \sum x^2 - n\bar{x}^2$.

• For data in a frequency table, $S_{xx} = \sum (x - \bar{x})^2 f = \sum x^2 f - \dfrac{\left(\sum xf\right)^2}{n} = \sum x^2 f - n\bar{x}^2$.

• Variance, $s^2 = \dfrac{S_{xx}}{n-1}$.

• Standard deviation, $s = \sqrt{\text{variance}} = \sqrt{\dfrac{S_{xx}}{n-1}}$.

• A data item may be considered to be an outlier if it is more than two standard deviations away from the mean.

The need for a measure of dispersion

Imagine you are deciding where to go on holiday in July. You have narrowed it down to two destinations, each with an average July temperature of 27°C. Knowing that one destination has a minimum July temperature of 21°C and a maximum of 32°C while the other has a minimum of 10°C and a maximum of 41°C would provide you with useful additional information.

The range

The range is the simplest measure of spread; it is calculated by subtracting the smallest data value from the largest data value. For the two holiday destinations above, the ranges would be:

$32 - 21 = 11°C$ and $41 - 10 = 31°C$.

Range involves only two data points and so can be unrepresentative, particularly if there is at least one outlier and so the value of one or both of those data points is very large or very small. Other measures, like interquartile range, are less prone to this. This section covers variance and standard deviation and the calculation of these measures which is based on the whole data set and how far the data values are from the mean.

Common mistake: Be careful not to confuse the range and the mid-range. The range is a measure of spread but the mid-range is an average.

The sum of squares (of deviations from the mean)

Variance and standard deviation, considered in the next sections, use the sum of squares of deviations from the mean.

$$S_{xx} = \sum (x - \bar{x})^2 = \sum x^2 - \frac{(\sum x)^2}{n} = \sum x^2 - n\bar{x}^2.$$ The first format of the formula makes it easier to understand what is being calculated but the second and third involves less work when doing the calculation. $(x - \bar{x})$ is the distance of a data item from the mean. Some data items will be above the mean, others will be below it; squaring $(x - \bar{x})$ makes all these values positive. The measures of dispersion that use S_{xx} are measuring how far all the data items are from the mean. If they are all near the mean, the data are not very spread out.

Worked example

Example 1

Eight students are playing a game with dice. They throw one of them and it needs to come up 6 before they can start. The following data show how many throws each of them takes to be able to start.

3   6   1   4   1   12   4   7

Calculate S_{xx}.

Solution

$$\sum x = 3 + 6 + 1 + 4 + 1 + 12 + 4 + 7 = 38$$

$$\sum x^2 = 3^2 + 6^2 + 1^2 + 4^2 + 1^2 + 12^2 + 4^2 + 7^2 = 272$$

$$S_{xx} = \sum x^2 - \frac{(\sum x)^2}{n} = 272 - \frac{38^2}{8}$$

$$S_{xx} = 91.5$$

First find the sum of the data values.

Next find the sum of the squares of the data values.

n is the number of data values (8 in this example).

Using this form of the formula makes calculation easy; using the form with $\bar{x}$ in will mean that you might end up working with long decimals. The formulae are given at the start of the exam paper.

Hint: You could just type the data into your calculator to get values of $\Sigma x, \Sigma x^2$ and n. The working is only shown here so that you are reminded of what your calculator is finding.

Common mistake: Σx^2 means 'square each data value and add them up'. Do not confuse it with $(\Sigma x)^2$ which would mean 'add the data values and square the answer'. These do not give the same result.

Hint: Your calculator will give you standard deviation automatically from a list of data or a frequency table.

Variance and standard deviation

For the above example, $s^2 = \dfrac{S_{xx}}{n-1} = \dfrac{91.5}{7} = 13.071\ 428\ ...$

Standard deviation, $s = \sqrt{\text{variance}} = \sqrt{13.071\ 428\ ...} = 3.62$ (2 d.p.).

	1–Variable
$\bar{x}$	4.75
$\sum x$	38
$\sum x^2$	272
σx	3.38193731
sx	3.61544306
n	8 $\downarrow$

Common mistake: Look at this screen shot. There are two values shown, each of which is sometimes referred to as standard deviation: the symbols are σx and sx on this calculator. The value you need for standard deviation is sx on this calculator screen; it might have different symbols on different calculators. The value you need is calculated using $\sqrt{\dfrac{S_{xx}}{n-1}}$; the other one is calculated using $\sqrt{\dfrac{S_{xx}}{n}}$ - it is a measure of spread sometimes used in introductory statistical work. The one you want is the larger of the two values.

Working with summaries of data

Sometimes in examination questions you will be given a summary of the data rather than a list of all the data or a frequency table.

You need to be able to use the formulae for mean and standard deviation when given summary data.

> ### Worked example
>
> #### Example 2
>
> 70 dice are rolled and the scores are noted.
>
> For these data, $n = 70$, $\sum x = 274$, $\sum x^2 = 1286$. Find the sample mean and sample standard deviation.
>
> #### Solution
>
> $\bar{x} = \dfrac{274}{70} = 3.914\ 285\ ...$
>
> $S_{xx} = \sum x^2 - \dfrac{\left(\sum x\right)^2}{n} = 1286 - \dfrac{274^2}{70}$
>
> $S_{xx} = 213.185\ 714\ ...$
>
> $s = \sqrt{\dfrac{S_{xx}}{n-1}} = \sqrt{\dfrac{213.485\ 714\ ...}{69}}$
>
> $s = \sqrt{3.093\ 995\ ...} = 1.758\ 975\ ...$
>
> $s = 1.76$ (3 s.f.)

Hint: Calculators usually have more than one memory. Use the memories to keep exact values of $\bar{x}$, S_{xx} etc. so that you can used them in later calculations.

Work out the mean first; remember you may need the unrounded answer for the rest of your working so put it in a calculator memory.

Work out S_{xx} before working out the standard deviation.

Common mistakes: Make sure you divide by 69 here, not 70.

Hint: Do not round until you have got your final answer. You should work with the unrounded value of the mean. If you are asked to work out variance then round it sensibly but if the variance is part of your working in finding standard deviation then wait until you have worked that out before rounding, otherwise your answer will be less accurate.

Working from a frequency table

The way you enter data into a calculator to calculate standard deviation from a frequency table is the same as for mean from a frequency table; your calculator should give you both mean and standard deviation at the same time.

Worked example

Example 3

The table below shows the number of goals scored by a football team in some of its matches. Find the sample mean and sample standard deviation.

Number of goals	0	1	2	3
Frequency	9	17	10	2

	List 1	List 2	List 3	List 4
SUB	goals	freq		
2	1	17		
3	2	10		
4	3	2		
5				

| 1-VAR | 2-VAR | REG | | SET |

Hint: Your calculator should allow you to type in the data as a frequency table. Make sure you know how to do this for your calculator.

Solution

The sample mean is 1.13 (to 2 d.p.).

	1–Variable
$\bar{x}$	1.13157894
Σx	43
Σx^2	75
σx	0.83259431
sx	0.84377057
n	38 ↓

The sample standard deviation is 0.844 (3 s.f.)

Common mistake: If your calculator uses lists make sure that the frequency is set to the correct list.

1 Var	XList	: List 1
1 Var	Freq	: List 2

Hint: Make sure you write down the correct value from the list of statistics in your calculator. Round the answer sensibly. You should also check that the value of n (the number of data values) is what you expect – in this case there were 38 matches; this is the total frequency.

Working from a grouped frequency table

This is the same as for an ungrouped frequency table but you need to find the midpoint of each group first.

Hint: Remember the standard deviation you want is the larger of the two values given by your calculator

Outliers

Outliers are unusually large or unusually small data values. They can occur for a number of reasons; they might be mistakes or they might be genuine data values that happen to be very large or small. There are several rules of thumb for identifying outliers.

A data item may be considered to be an outlier if it is more than two standard deviations away from the mean.

Common mistake: Remember that the mean and standard deviation you get by using midpoints for a grouped frequency table are only estimates.

Worked example

Example 4

A class are trialling a test. They time how long it takes them to complete it. The mean time is 69.4 minutes and the standard deviation is 11.3 minutes. How high, or low, would a student's time have to be for it to be considered an outlier?

Solution

$$\bar{x} + 2s = 69.4 + 2 \times 11.3 = 92$$
$$\bar{x} - 2s = 69.4 - 2 \times 11.3 = 46.8$$

Times which are above 92 minutes, or below 46.8 minutes could be considered outliers.

Hints: Sometimes it is obvious that a data item is a mistake. For example, a person 2.75 m tall would be taller than the world's tallest man so such a data item would be incorrect; it should either be ignored or, if possible, corrected. On the other hand, a data item giving the height of a person as 2.21 m could be genuine but, if possible, should be checked.

Test yourself

TESTED

1 The number of goals scored by a football team in a sample of games is:

2   3   5   7   7   1   2

Find S_{xx} for this sample (some of these answers have been rounded).

A   34.53   B   36.857   C   19.5   D   141   E   2.478

2 A class of students sat a test. Their mean mark was 63 and the standard deviation was not zero. Another student sits the test later and gains a mark of 63. The mean and standard deviation are recalculated to include this mark. Four of the following statements are false and one is true. Which one is true?

A   The mean stays the same but there is not enough information to say what happens to the standard deviation

B   There is not enough information to say what happens to the mean or the standard deviation.

C   The new mean is the same as before but the standard deviation is lower.

D   Both the mean and the standard deviation stay the same.

E   Both the mean and the standard deviation are decreased because of the additional mark.

7 coins are tossed by each of 20 students and the number of heads for each student is noted.

$\sum x = 68, \sum x^2 = 276$. Questions 3 and 4 use these data.

3 Find the mean number of heads obtained on the seven coins.

A   2.86 (3 s.f.)   B   3.4   C   4.06 (3 s.f.)   D   9.71 (3 s.f.)   E   0.294 (3 s.f.)

4 Find the variance of the number of heads on the coins.

A   1.50 (3 s.f.)   B   1.54 (3 s.f.)   C   2.24   D   2.36 (3 s.f.)   E   14.53 (2 d.p.)

5 A sample of married women in their forties are asked how many children they have had.

Number of children	0	1	2	3	4	5	6
Frequency	13	17	38	20	8	3	1

Four of the following statements are false and one is true. Which one is true?

A $S_{xx} = 588$.

B   There were 206 women in the sample.

C $n = 7$.

D $S_{xx} = 159.35$ (2 d.p.).

E   The variance of children per woman was 1.653 (3 d.p.).

Full worked solutions online

CHECKED ANSWERS

Exam-style question

A bakery makes loaves of bread labelled 400 grams. A sample of ten loaves is taken and their weights, in grams, are:

| 407.0 | 397.2 | 412.4 | 422.8 | 422.0 | 427.3 | 388.2 | 407.7 | 421.0 | 399.1 |

i   For this sample find
  ● the sample mean,
  ● the sample standard deviation.

ii For the next sample of ten loaves, $\sum x = 3935.7$ and $\sum x^2 = 1\,598\,296.4$.
   For this sample, find
  ● the sample mean,
  ● the sample standard deviation.

iii Comment on your answers for the two samples.

Short answers on page 158

Full worked solutions online

CHECKED ANSWERS

Bivariate data

REVISED

Key facts

● Bivariate data are in pairs so they involve two variables, e.g. age and height of a sample of children. Such data can be plotted on a scatter diagram.

● The **independent** variable goes on the x-axis of a scatter diagram and the **dependent** variable goes on the y-axis.

● An outlier in a scatter diagram is a point that does not fit the general pattern of the rest of the data.

● Correlation means a linear relationship. It is possible to measure the strength of a linear relationship.

● Even if there is strong correlation, it does not mean that change in one of the variables **causes** change in the other variable.

● A scatter diagram might reveal a relationship between the variables which is not linear; association means any relationship between the variables.

● A data set which consists of values of a variable together with the times at which they occurred is a time series; the time is plotted on the horizontal axis. If the dependent variable varies continuously with time, the points may be joined (dot to dot) to show the trend.

Patterns in scatter diagrams

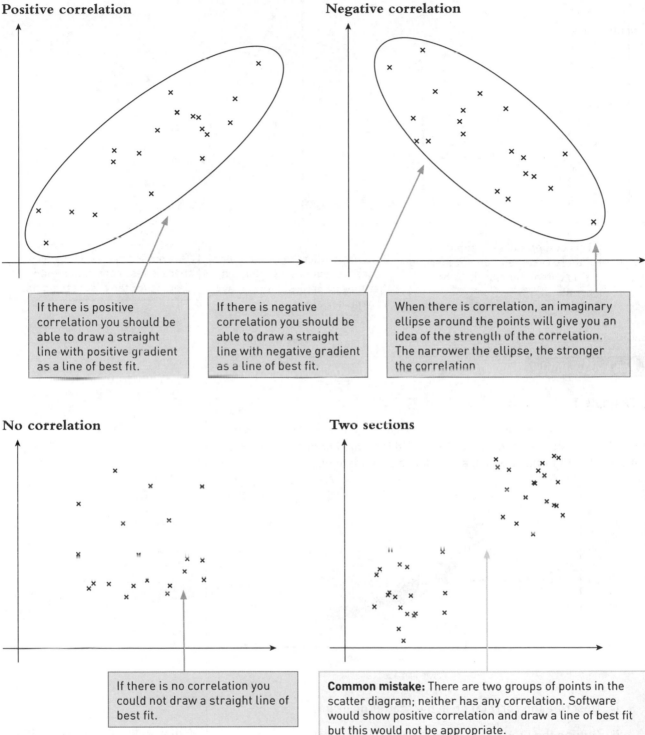

Positive correlation

If there is positive correlation you should be able to draw a straight line with positive gradient as a line of best fit.

Negative correlation

If there is negative correlation you should be able to draw a straight line with negative gradient as a line of best fit.

When there is correlation, an imaginary ellipse around the points will give you an idea of the strength of the correlation. The narrower the ellipse, the stronger the correlation

No correlation

If there is no correlation you could not draw a straight line of best fit.

Two sections

Common mistake: There are two groups of points in the scatter diagram; neither has any correlation. Software would show positive correlation and draw a line of best fit but this would not be appropriate.

Outliers

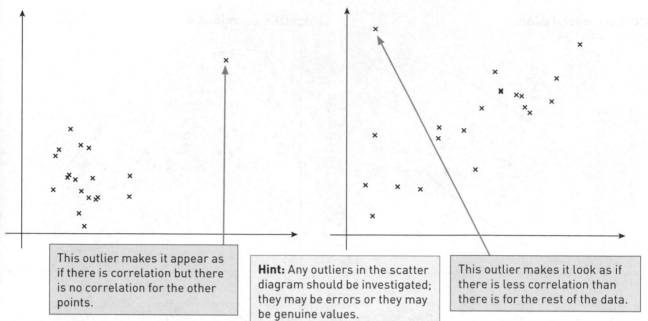

This outlier makes it appear as if there is correlation but there is no correlation for the other points.

Hint: Any outliers in the scatter diagram should be investigated; they may be errors or they may be genuine values.

This outlier makes it look as if there is less correlation than there is for the rest of the data.

Worked example

Example 1

The scatter diagram below shows the median age and the birth rate per 1000 in all the countries of the world for which data are available. A line of best fit has been drawn using a spreadsheet.

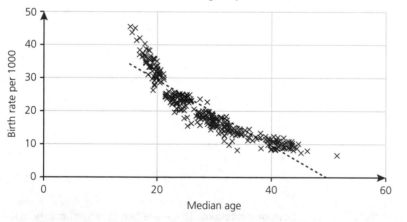

i   Describe the correlation in the scatter diagram.

ii   Outline the limitations of the straight line model.

iii   Sketch the graph of a better model.

Hint: Although the straight line model could be improved, all points do lie fairly close to the line and the line has a negative gradient so there is strong negative correlation.

Solution

i   *There is strong negative correlation.*

ii   *For countries with a high or low median age, the straight line suggests a lower birth rate than the actual birth rate but for countries with a medium median age, the line suggests a birth rate that is generally too high.*

Hint: Look at different sections of the data set where the straight line model could be improved.

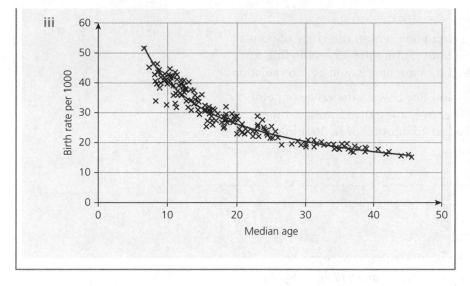

Hint: A curved model is better.

Common mistake: Do not continue the model beyond the range of the data; it might not be appropriate. In this case, there are no countries with higher or lower median ages than those shown.

Using best fit models

The scatter diagram below shows the weight of strawberries picked by a fruit picker against the time taken. Readings of the weight of fruit picked were taken every two minutes.

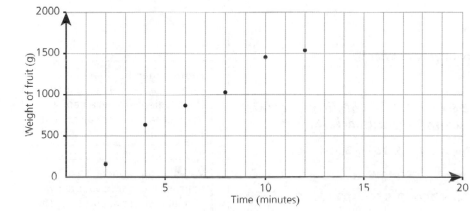

Hint: The independent variable goes on the *x*-axis. In this case, the experimenter is interested in how weight of fruit picked depends on time so time is the **independent variable** and weight of fruit picked is the **dependent variable**.

It looks as though a linear relationship would describe the data quite well.

If a straight line is drawn to show the relationship, it cannot go through all the points because the correlation is not perfect.

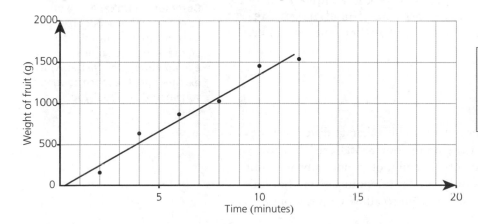

When a line of best fit is calculated (for example using software or calculator functions) it is known as a **regression line**.

Interpolation and extrapolation

Interpolation is estimating for a data point within the range of points you already have. For the example above, using the regression line to estimate the amount of fruit picked in 5 minutes would be interpolation.

Extrapolation is using the regression line beyond the range of points you already have. For the example above, using the regression line to estimate the amount of fruit picked in 15 minutes or in 30 minutes would each be examples of extrapolation.

Worked example

Example 2

The equation of the line of best fit for the fruit picker graph is $y = 136x - 3.85$, where x is the time in minutes and y is the weight of fruit in grams.

  i Give an interpretation of the gradient of the line.

  ii Estimate the weight of fruit the picker would pick in

    a 15 minutes,

    b 30 minutes.

  iii Comment on the reliability of your answers.

Solution

  i The gradient of $y = 136x - 3.85$ is 136; it represents the weight of fruit (in grams) picked each minute.

  ii a $x = 15 \Rightarrow y = 136 \times 15 - 3.85 = 2036.15$.

      2040 g (3 s.f.)

    b $x = 30 \Rightarrow y = 136 \times 30 - 3.85 = 4076.15$

      4080 g (3 s.f.)

  iii The answer for 15 minutes is likely to be fairly reliable because 15 is near to the data in the scatter diagram. The answer for 30 minutes will be less reliable as it is further from the data; it assumes the straight line relationship will continue but the picker may get tired and slow down with time.

> **Hint:** Remember that in the form $y = mx + c$, m represents the gradient. The gradient is the increase in y for a one unit increase in x.

> When using a model, it isn't appropriate to give your answer to a lot of decimal places; in this case the number of grams the picker would pick in a given time would probably vary quite a bit.

> **Common mistake:** A line of best fit or other model is often used for making predictions so there may be some extrapolation; the further you extrapolate, the more cautious you should be about your answer.

Time series

A time series consists of the values of a variable taken at different times. The values are plotted with the time on the horizontal axis. To show the trend over time, the points can be joined (dot to dot); a broken line should be used if the data values only exist at the points given. Several time series can be plotted on the same graph to allow comparison.

The graph below shows the height of a river in Texas from 1 August to 29 August 2017.

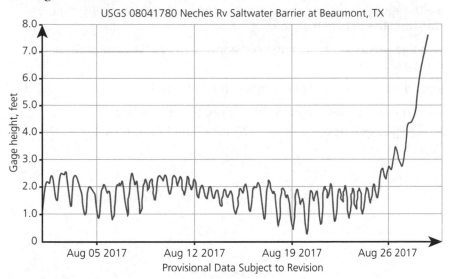

USGS 08041780 Neches Rv Saltwater Barrier at Beaumont, TX

Provisional Data Subject to Revision

Worked example

Example 3

Make two comments about the trends shown in the height of the river.

Solution

The height of the river varies over the course of a day.

From 26 August onwards the river is getting gradually higher; it becomes much higher than usual.

Hint: The question asks you to make two comments. There are more than two comments which could be made in this context but, in an examination, you should give the number of comments asked for.

Test yourself

TESTED ☐

1 Fred exercises for 30 minutes then observes his pulse rate after stopping the exercise. The data are shown in the table and scatter diagram below.

Time after exercise (min), x	Pulse rate, y
½	133
1	127
1½	98
2	67
2½	68

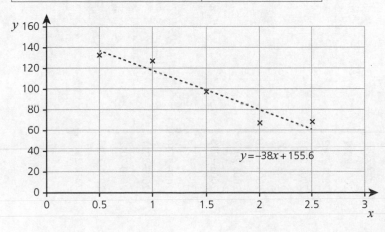

$y = -38x + 155.6$

Three of the following statements are false and one is true. Find the one that is true.

A   It would be reasonable to use the equation of the regression line to find the pulse rate 4 minutes after exercise as this is not far from the data in the table.

B   The equation of the regression line which you have found will apply to the pulse rate of all men who exercise for 30 minutes then stop.

C   Time after exercise is the dependent variable and pulse rate is the independent variable.

D   79.6 is the model's prediction of the pulse rate 2 minutes after Fred stops exercise.

2   The scatter diagram below has the regression line drawn on it. Which of the following could be the equation of the regression line?

A $y = -1.1x - 3.3$ B $y = -1.1x + 3.3$ C $y = 1.1x - 3.3$ D $y = 1.1x + 3.3$

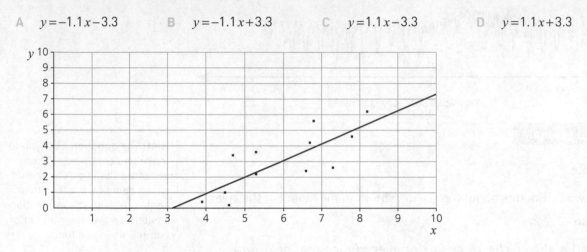

3   The graph below shows male life expectancy at birth for four regions of England. Values are estimates over three years at a time.

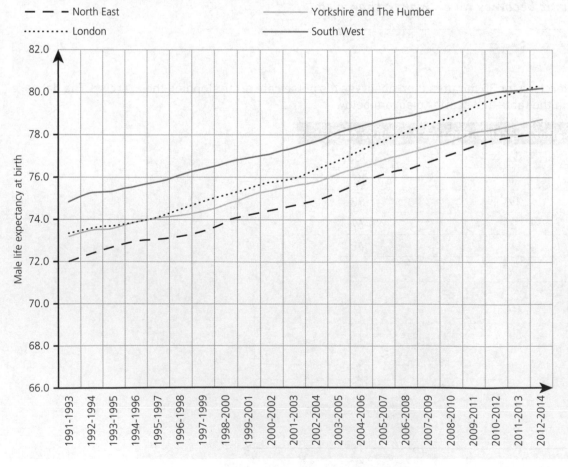

Four of the following statements are false and one is true. Find the one that is true.

A   Male life expectancy in the South West has increased by 5 years over the course of about 20 years and this trend will continue into the future.

B   Male life expectancy in London has increased more rapidly than in the other three areas.

C   The gap between male life expectancy in the North East and the South West has remained constant throughout the time period covered by the graph.

D   The graph would show the trends better if the vertical axis started at zero.

E   Male life expectancy in Yorkshire and The Humber has risen slower than in the other three regions.

4   The scatter diagram below shows unemployment rates and life expectancy at birth for the countries of Europe.

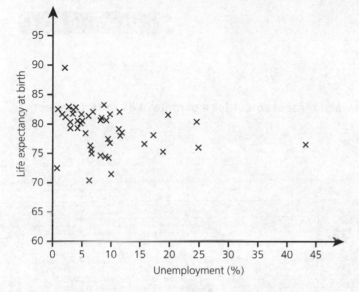

Three of these statements are false and one is true. Find the one that is true.

A   Living in a country with high unemployment causes lower life expectancy.

B   The unemployment rate between 40% and 45% is an outlier and it must be an error.

C   The life expectancy of nearly 90 years is unrealistically high so it must be an error.

D   There is weak negative correlation between unemployment and life expectancy.

5   The graph below shows the number of marriages per 1000 unmarried men and per 1000 unmarried women in England and Wales.

Marriages per 1000 unmarried people aged 16 and over

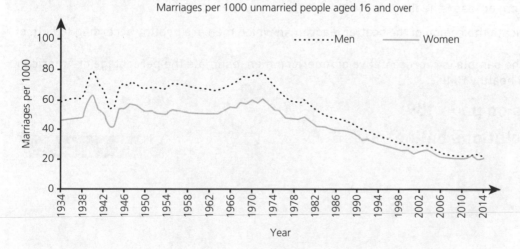

Four of the following statements are true and one is false. Find the one that is false.

A The graph shows that in 2014 about one in every 50 unmarried adults got married.

B The graph shows that more men than women get married.

C It is possible to infer that, for the time period shown by the graph, there were more unmarried women than unmarried men in England and Wales.

D If a scatter diagram was plotted with marriages per 1000 men on the horizontal axis and marriages per 1000 women on the vertical axis then there would be very strong positive correlation.

E Marriages per 1000 unmarried people shows the trend in marriages better than total marriages in the population as it eliminates the effect of population change on the statistics.

Full worked solutions online

CHECKED ANSWERS

Exam-style question

The scatter diagram below shows the BMI and waist measurement for a sample of 89 American men.

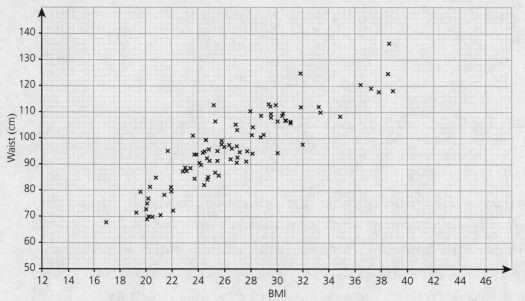

i Describe the correlation between BMI and waist measurement.

A website says that a healthy BMI for adults is in the range 19 to 25. Another website says that a waist measurement of 94 cm or less is healthy for men.

ii On a sketch, indicate the region of the scatter diagram in which men are healthy according to both of the websites.

iii Assuming that the sample is representative of American men, estimate the percentage of American men who have a healthy BMI.

Short answers on page 159

Full worked solutions online

CHECKED ANSWERS

Chapter 3 Probability

About this topic

Although the basic ideas of probability can seem quite straightforward, it is important to think clearly to avoid going wrong. Probability is used in risk assessment so a good understanding can be very useful in all kinds of situations.

Probability distributions are used for discrete random variables. A random variable can take different values. A discrete random variable can only take discrete values. Examples of discrete random variables include the total score when two dice are thrown, the proportion of heads when tossing 4 coins, the number of children in a family. The probability distribution gives the probability of each possible value.

Before you start, remember ...

- Basic ideas of probability and calculations with fractions and decimals from GCSE.
- Types of data and vertical line charts from Chapter 2.
- Substituting into formulae from GCSE.

Working with probability

REVISED

> ### Key facts
>
> - The probability of an event A can often be found using
> $$P(A) = \frac{\text{Number of ways } A \text{ can occur}}{\text{Total number of outcomes}}.$$ This only works if all outcomes are equally likely.
> - The expected frequency of event A in n trials is $nP(A)$.
> - $P(A') = 1 - P(A)$ where A' is the event 'not A'.
> - A probability of zero means an event cannot happen; a probability of 1 means it is certain to happen.
> - Venn diagrams can be used to show either the number of outcomes or the probabilities.
> - For mutually exclusive events, $P(A \text{ or } B) = P(A) + P(B)$.
> - For independent events, $P(A \text{ and } B) = P(A) \times P(B)$.
> - Sample space diagrams are used to display the outcomes when you have two trials happening together, each with equally likely outcomes.
>
> | | H | T |
> |---|----|----|
> | **H** | HH | HT |
> | **T** | TH | TT |

- Tree diagrams can be used for working out the probability of two (or more) events.

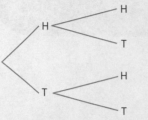

- Once you have the correct probabilities on a tree diagram, you multiply along the branches to find the probability that all the relevant events happen.
- P(*A* happens at least once) = 1 − P(*A* never happens).

Finding probabilities using equally likely outcomes

You will almost certainly have used the following method to find probabilities:

$$P(A) = \frac{\text{Number of ways } A \text{ can occur}}{\text{Total number of outcomes}}.$$

> **Common mistake:** This only works if all outcomes are equally likely. It can be used in quite complicated situations, but you must be careful to count the equally likely outcomes correctly.

Worked example

Example 1

What is the probability of this spinner landing on red, assuming that it is fair?

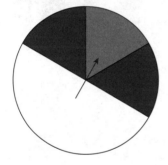

Solution

The sections are not all equal but you can put in some imaginary lines to get equal sections.

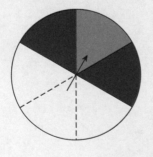

$Probability = \frac{2}{6} = \frac{1}{3}.$

> You could also do this question by looking at the fraction of the total area that is coloured red. There is sometimes more than one correct way to approach a probability question; practice will help you find the ways that you prefer.

> Give answers to probability questions as fractions or decimals.

Estimating probability

For the spinner from Example 1, an experiment was carried out to estimate the probability that the spinner lands on red.

The spinner was spun 1000 times and the numbers of times it came up red were used to give progressive estimates of the probability using

$$\text{Estimated probability} = \frac{\text{Number of times the spinner lands on red}}{\text{Total number of trials}}.$$

The results are shown on this graph.

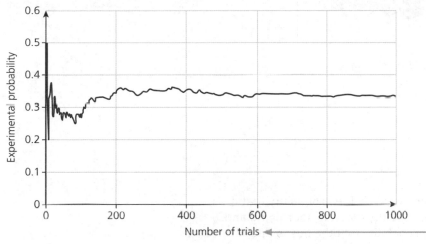

In this case, a trial consists of spinning the spinner once.
A trial is an experiment.

The graph shows the **experimental** estimates of the probability.

You have already seen in Example 1 that the **theoretical** probability of getting red is $\frac{1}{3}$.

If the experimental probability, based like this on a large number of trials, had turned out to be different from the theoretical, there would have been evidence that the spinner was **biased**.

Expected frequency of an event

In 1000 trails of spinning the fair spinner from Example 1, the expected number of times red appears is $1000 \times \frac{1}{3} = 333\frac{1}{3}$.

Impossible or certain events

- If an event cannot happen, its probability is 0.
- If an event is bound to happen, its probability is 1.

The probability that an event does not happen

Either an event happens or it does not. $P(A') = 1 - P(A)$ where A' is the event 'not A'.

> **Common mistake:** The expected frequency is the average number of times you would get red if you repeated the 1000 trials many times; it does not mean that you would be surprised if you did not get red $333\frac{1}{3}$ times. Since the expected frequency is an average, it need not be a whole number.

Worked example

Example 2

The weather forecast estimates the probability that it will rain tomorrow as $\frac{1}{4}$. What is the probability that it will not rain?

Solution

$P(no\ rain) = 1 - \frac{1}{4} = \frac{3}{4}$.

Using Venn diagrams

Venn diagrams are useful for showing the relationship between two, or more, events. They can be very helpful for finding probabilities.

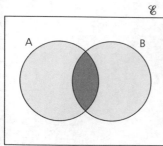 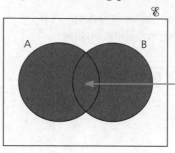

Notice that $A \cup B$ includes the possibility that both of A and B could happen.

$A \cap B$ 'A and B' $A \cup B$ 'A or B'

Venn diagrams for probability

Worked example

Example 3

The probability that it rains on a given day is 0.7, the probability that it rains and the bus is late is 0.4. The probability that it is not raining and the bus is not late is 0.1. What is the probability that the bus is late?

Solution

Let R denote the event that it rains, B denote the event that the bus is late.

1 First fill in the probability of both events, in the middle region.

2 The probability that neither event occurs is 0.1.

3 The probability in this region needs to add to the 0.4 in the middle to give 0.7.

4 All the probabilities need to add up to 1 so the probability for this region is 0.2.

0.1

0.4

0.3   0.2

R   B

P(bus late) = 0.4 + 0.2 = 0.6

Venn diagrams can also be used for showing the number of outcomes, or the numbers in particular sets.

Venn diagrams for number of objects

Worked example

Example 4

There are 21 cars in a showroom, 13 of them are silver. 4 of the silver cars are worth over £20000. Two of the cars are neither silver nor worth over £20000. One of the cars in the showroom is chosen at random as a prize in a competition. What is the probability that it is worth over £20000?

Solution

Let S denote the set of silver cars, V denote the set of cars worth over £20000.

1 You know the number of cars in the intersection so fill this in first.

2 The total number of silver cars is 13 so there are 9 cars in this region.

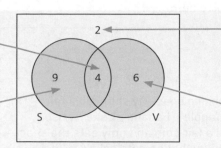

3 There are 2 cars which are neither silver nor worth over £20 000.

4 The total number of cars is 21 so there are 6 cars in this region.

The number of cars worth over £20000 is 4 + 6 = 10.

P(worth over £20 000) = $\frac{10}{21}$.

Mutually exclusive events

Mutually exclusive events cannot happen together.

A Venn diagram for two mutually exclusive events is shown below.

For mutually exclusive events, P(A or B) = P(A) + P(B).

Common mistake: Be careful not to add probabilities automatically when you want the probability of one event *or* another event; this only works when they are mutually exclusive.

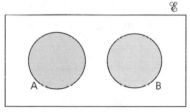

Sample space diagrams

A sample space diagram is a table showing all the possible outcomes from two trials (or experiments), each of which has equally likely outcomes.

Worked example

Example 5

A fair spinner, with numbers 1 to 3 on it, is spun and a calculator chooses a random integer between 1 and 6 (inclusive). The two numbers are multiplied to give a score. What is the most likely score and what is the probability of getting it?

Solution

Hint: If the question involves two numbers being multiplied or added etc. then it makes your working easier to write the relevant results in the cells in the sample space.

	Score		Number on calculator					
			1	2	3	4	5	6
Number		1	1	2	3	4	5	6
on		2	2	4	6	8	10	12
spinner		3	3	6	9	12	15	18

This cell in the sample space represents the event '2 on the spinner and 4 on the calculator'.

The most frequently occurring score is 6 with a probability of $\frac{3}{18} = \frac{1}{6}$.

Sample space diagrams need not involve numbers.

Worked example

Example 6

Christopher decides on his mid-morning snack by spinning a fair spinner which he has labelled with 'apple', 'banana', 'cake', 'crisps', 'chocolate', 'raisins'. He then tosses a fair coin and only eats the snack if it lands heads. What is the probability that he eats cake?

Solution

	Apple	Banana	Cake	Crisps	Chocolate	Raisins
Head (eats)						
Tail (does not eat)						

Each cell in the sample space represents an equally likely outcome.

Probability of eating cake $= \dfrac{1}{12}$.

Independent events

In Example 6, the result on the spinner and the result on the coin are independent events; neither of them affects the other. The probability that Christopher eats cake can be calculated using $P(A \text{ and } B) = P(A) \times P(B)$.

P(spinner shows 'cake' and coin shows heads) $= \dfrac{1}{6} \times \dfrac{1}{2} = \dfrac{1}{12}$.

Tree diagrams

Tree diagrams are useful for showing the possible things that can happen and also for working out probabilities of two (or more) events.

Worked examples

Example 7

A box contains three red beads and two gold beads. Apart from the colour, the beads are identical. A contestant on a game show is blindfolded. She pulls out a bead then puts it back. She then pulls out a second bead. She wins a prize if she pulls out exactly one gold bead. What is the probability that she wins a prize?

Solution

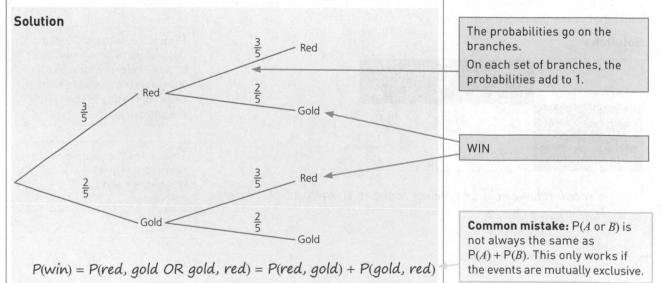

The probabilities go on the branches.

On each set of branches, the probabilities add to 1.

WIN

$P(\text{win}) = P(\text{red, gold OR gold, red}) = P(\text{red, gold}) + P(\text{gold, red})$

Common mistake: $P(A \text{ or } B)$ is not always the same as $P(A) + P(B)$. This only works if the events are mutually exclusive.

$$P(\text{win}) = \frac{3}{5} \times \frac{2}{5} + \frac{2}{5} \times \frac{3}{5}$$
$$= \frac{6}{25} + \frac{6}{25} = \frac{12}{25}$$

Hint: Working along the branches, multiply the probabilities to find the probability that both gold and red occur.

A tree diagram can be as big as you need it to be.

Worked example

Example 8

My calculator gives me a random number with 3 decimal places. Assuming that each digit is equally likely to be from 0 to 9 inclusive, what is the probability that a random number contains at least one digit 7?

Solution

There are ten equally likely possibilities for the first digit so $P(7) = 0.1$.

These are the possibilities for the first digit. We are only interested in whether or not it is 7.

$$P(\text{at least one 7}) = 1 - P(\text{no 7s})$$
$$= 1 - 0.9 \times 0.9 \times 0.9 = 0.271$$

Hint: This is the easiest way to work out P(at least one 7).

Test yourself

TESTED

1 Rosy and Julian are taking it in turns to spin a fair spinner with the numbers 1 to 6 on it. So far it has landed 6 1 4 6 2 2 2 4
Four of the following statements are false and one is true. Find the one that is true.

A The probability of it showing 2 on the next throw is $\frac{3}{8}$.

B It has not shown 5 yet so there is a high probability of getting 5 on the next throw.

C Getting three 2s in a row is very unlikely so there must be a typing error in the results.

D If they keep spinning the spinner, roughly one sixth of the time the spinner will show 3.

E This spinner is more likely to land showing an even number than an odd number.

2 Raffle tickets numbered 277 to 575 inclusive are put into a large container and one ticket is taken out at random. What is the probability of a number divisible by 5 being chosen?

A $\frac{1}{5}$ B $\frac{30}{149}$ C $\frac{60}{299}$ D $\frac{59}{298}$ E $\frac{59}{299}$

3 Some gardeners have phoned in to ask for their garden to appear on TV. The number of these gardens having certain features is shown in the Venn diagram below. One of these gardens will be chosen at random to be in the TV show. What is the probability that a garden with both a vegetable patch and a pond will be chosen?

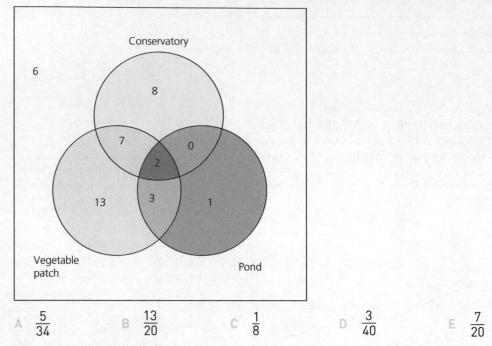

A $\frac{5}{34}$ B $\frac{13}{20}$ C $\frac{1}{8}$ D $\frac{3}{40}$ E $\frac{7}{20}$

4 A box contains 50 chocolates. 20 of them are coated with dark chocolate, the rest with white chocolate. 28 chocolates are toffee-centred, the rest fudge. Of the dark chocolates, 12 have toffee centres. Alix chooses a chocolate at random. Four of these statements are true and one is false. Find the one that is false.

A The probability of choosing one with a toffee centre and dark chocolate coating is 0.6.

B The probability of choosing one that does not have a dark chocolate coating is 0.6.

C The probability of choosing one with a toffee centre or a dark chocolate coating is 0.72.

D The probability of choosing one which does not have a toffee centre or a dark chocolate coating is 0.28.

E More than half of the white chocolates have toffee in them.

5 A jar contains six white discs and four black discs. A disc is taken out at random and replaced then the process is repeated. What is the probability of getting a black disc at least once?

A $\frac{12}{25}$ B $\frac{16}{25}$ C $\frac{2}{3}$ D $\frac{3}{4}$ E $\frac{2}{5}$

Full worked solutions online

CHECKED ANSWERS

Exam-style question

A new £1 coin was introduced in the UK in 2017. At the time of introduction, it was estimated that 3% of the old £1 coins were fake. You should assume that fake coins occur independently of each other with 3% probability.

i A money box contains 400 old £1 coins. What is the expected number of fake coins?

ii Alice has three old £1 coins. What is the probability that none of them are fake?

iii What is the minimum number of old £1 coins that someone would need to have before the probability of having at least one fake is above 0.5?

iv Give an example of circumstances when it would not be reasonable to assume that fake coins occur independently of each other with 3% probability.

Short answers on page 159

Full worked solutions online

CHECKED ANSWERS

Probability distributions

> **Key facts**
>
> - Upper case (capital) letters are used to stand for discrete random variables; e.g. S could stand for 'the total score when two dice are thrown'.
> - Lower case letters are used to stand for values of the discrete random variable; e.g. $s = 2, 3, 4, ..., 12$.
> - Each possible value of the random variable has a probability of occurring which is between 0 and 1.
> - If the random variable X can take values $r_1, r_2, r_3, ..., r_n$ with probabilities $p_1, p_2, p_3, ..., p_n$ respectively then $\sum_{i=1}^{n} p_i = 1$, i.e. all the probabilities add up to 1.
> - A discrete probability distribution can be illustrated by a vertical line chart.

Discrete random variables

Discrete random variables can only take discrete values. Each possible value has a probability of occurring.

> **Worked example**
>
> ### Example 1
>
> A fair coin is tossed twice. Fill in the probability of each possible number of heads.
>
> | Number of heads | 0 | 1 | 2 |
> |---|---|---|---|
> | Probability | | | |
>
> **Solution**
>
> The probabilities can be found by drawing a sample space diagram (shown on the right).
>
> | Number of heads | | First toss | |
> |---|---|---|---|
> | | | H | T |
> | Second toss | H | 2 | 1 |
> | | T | 1 | 0 |
>
> There are four items in the sample space, one of them is '0 heads'.
>
> | Number of heads | 0 | 1 | 2 |
> |---|---|---|---|
> | Probability | $\frac{1}{4}$ | $\frac{1}{2}$ | $\frac{1}{4}$ |

The probability distribution of a discrete random variable

The probability distribution can be a table which gives a probability for each value (as in Example 1 above) or it can be a formula which gives a probability for each value (as in Example 2 below).

Worked examples

Example 2

The random variable, X, can take values 1, 2, 3, 4. The probability of each possible value is given by $P(X = r) = \dfrac{5}{4r(r+1)}$. Find $P(X = 2)$.

Solution

$$P(X = 2) = \frac{5}{4 \times 2 \times 3} = \frac{5}{24}.$$

Put 2 in the formula in place of r.

Example 3

The random variable X has a probability distribution given by the formula $P(X = r) = k(7 - 2r)$ for $r = 1, 2, 3$. k is a constant. Find the value of k.

Solution

r	1	2	3
$P(X = r)$	$5k$	$3k$	k

The probabilities add up to 1 so $5k + 3k + k = 1$

$$9k = 1$$
$$k = \frac{1}{9}$$

- For each possible value of the random variable, its probability is between 0 and 1.
- If the random variable X can take values $r_1, r_2, r_3, \ldots r_n$ with probabilities $p_1, p_2, p_3, \ldots p_p$ respectively then $\displaystyle\sum_{i=1}^{n} p_i = 1$, i.e. all the probabilities add up to 1.

Hint: Start by putting the values of r into the formula and writing the probabilities in a table.

Hint: Knowing that all the probabilities add up to 1 will be useful for answering exam questions but it is also a useful check when working out all the probabilities in a probability distribution.

Illustrating discrete random variables

REVISED

A vertical line chart is a good way to illustrate the probability distribution of a discrete random variable. The following line chart illustrates the probability distribution in Example 3.

r	1	2	3
$P(X = r)$	$\dfrac{5}{9}$	$\dfrac{3}{9}$	$\dfrac{1}{9}$

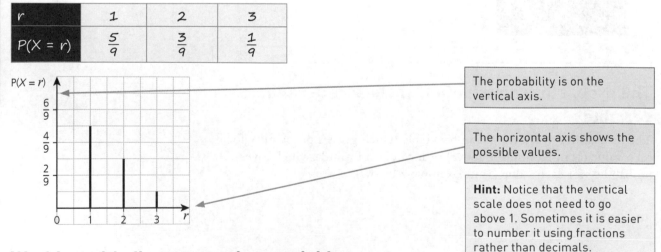

The probability is on the vertical axis.

The horizontal axis shows the possible values.

Hint: Notice that the vertical scale does not need to go above 1. Sometimes it is easier to number it using fractions rather than decimals.

Working with discrete random variables

You can use the probabilities for discrete random variables in all the usual ways. Different formulae can be used for the probabilities for different values.

Worked example

Example 4

The probability distribution of the random variable X is given by

$$P(X = r) = 0.2 \times 0.8^{r-1} \quad \text{for } r = 1, 2, 3, 4, 5$$
$$P(X = r) = k \quad \text{for } r = 6 \; (k \text{ is a constant})$$
$$P(X = r) = 0 \quad \text{otherwise}$$

> This is just a way of saying that values of r which are not 1, 2, 3, 4, 5, 6 cannot happen.

i Find the value of k.

ii Two independent values of X are generated, one after the other, find the probability that their total is greater than 10.

Solution

i Using the formula $P(X = r) = 0.2 \times 0.8^{r-1}$, for values of r from 1 to 5, gives the probabilities in this table:

r	1	2	3	4	5	6
$P(X = r)$	0.2	0.16	0.128	0.1024	0.08192	k

> The probability for 6 is just k.

> **Hint:** All the probabilities add up to 1.

$$0.2 + 0.16 + 0.128 + 0.1024 + 0.08192 + k = 1$$

$$0.67232 + k = 1 \text{ so } k = 0.32768$$

ii A total greater than 10 can come from these pairs of values: (5, 6) or (6, 5) or (6, 6)

> **Common mistake:** 5 then 6 is different from 6 then 5, so both need to be included.

$$P(\text{total} > 10) = P(5, 6) + P(6, 5) + P(6, 6)$$

$$P(\text{total} > 10) = 0.08192 \times 0.32768 + 0.32768 \times 0.08192$$
$$+ 0.32768 \times 0.32768 \approx 0.161$$

> The two values are independent so to work out the probability of 5 and 6 you multiply the two probabilities.

Modelling with discrete random variables

Theoretical probability distributions are used as approximations to real-life situations so that predictions can be made about what is likely to happen.

> **Hint:** You could draw a tree diagram for part ii using the probabilities from part i and you should do so if it will help you to see how to work it out.

Worked example

Example 5

The number of people travelling in a car on a motorway, X, is modelled by the probability distribution $P(X = r) = k \times 0.4^r$ for $r = 1, 2, 3, 4$. k is a constant.

i Find the proportion of cars which, according to the model, only have one person travelling in them.

ii Would you expect the model to be suitable for the number of people travelling in a car on a road leading to a holiday resort in the summer?

> **Hint:** You need to work out the value of k so start by putting the values of r into the formula and writing the probabilities in a table.

Solution

i

r	1	2	3	4
$P(X = r)$	0.4k	0.16k	0.064k	0.0256k

$$0.4k + 0.16k + 0.064k + 0.0256k = 1$$

> Remember all the probabilities add up to 1.

$$0.6496k = 1$$
$$k = 1 \div 0.6496 = 1.54 \text{ (3 s.f.)}$$
$$P(X = 1) = 0.4k = 0.4 \times 1.54 = 0.616.$$

About 62% of cars only have one person in them.

ii No because most cars going to a holiday resort in the summer would have more than one person in.

Hint: The final answer can be left as a decimal or given as a percentage but it makes sense to round it as the value of k was not exact.

Test yourself

1 One of the five options in this question could be the probability distribution of a discrete random variable. The other four are not. Find the one which is.

A

r	0	1	2	3
$P(X=r)$	0.56	k	0.24	0.21

B $P(X=r) = k(r+1)$ for $r = 0, 1, 2$

   $P(X=r) = kr^2$ for $r = 4, 5$

   $P(X=r) = 0$ otherwise

C $P(X=r) = k(r^2 - 2r)$ for $r = 1, 2, 3$

D

r	2	3	4	5
$P(X=r)$	0.36	0.18	0.23	0.21

E $P(X=r) = \dfrac{kr}{r+1}$ for all values of r between 1 and 5 (inclusive) i.e. $1 \le r \le 5$

2 A discrete random variable, X, has the probability distribution:

$P(X=r) = k(2r^2 - r)$ For $r = 1, 2, 3$

$P(X=r) = 3k$ For $r = 4$

$P(X=r) = 0$ Otherwise

Find the value of k.

A 25 B $\dfrac{1}{22}$ C $\dfrac{1}{54}$ D 1 E 0.04

3 A discrete random variable, X, has the probability distribution shown in the table:

r	0	1	2	3	4
$P(X=r)$	0.25	0.2	0.15	0.3	0.1

Find $P(1 \le X < 3)$.

A 0.15 B 0.35 C 0.45 D 0.55 E 0.65

4 $X =$ the larger score when two fair dice are thrown. Which of the following is the correct probability distribution for X?

A $P(X=r) = \dfrac{2r-1}{36}$ for $r = 1, 2, 3, 4, 5, 6$

B $P(X=r) = \dfrac{1}{6}$ for $r = 1, 2, 3, 4, 5, 6$

C $P(X=r) = \dfrac{2r+1}{36}$ for $r = 1, 2, 3, 4, 5, 6$

D $P(X=r) = \dfrac{r}{21}$ for $r = 1, 2, 3, 4, 5, 6$

E $P(X=r) = \dfrac{(r-1)}{15}$ for $r = 1, 2, 3, 4, 5, 6$

5 Barry Mitchell is a footballer who has a 0.75 chance of scoring a goal from a penalty. He takes two penalties. Assume that these are independent of each other. Which of the following could be the probability distribution of the number of goals he scores?

A

No of goals	0	1	2
Probability	$\frac{1}{16}$	$\frac{3}{16}$	$\frac{9}{16}$

B

No of goals	0	1	2
Probability	$\frac{1}{16}$	$\frac{3}{16}$	$\frac{12}{16}$

C

No of goals	0	1	2
Probability	0	0.75	0.5625

D

No of goals	0	1	2
Probability	$\frac{1}{3}$	$\frac{1}{3}$	$\frac{1}{3}$

E

No of goals	0	1	2
Probability	$\frac{1}{16}$	$\frac{3}{8}$	$\frac{9}{16}$

Full worked solutions online

CHECKED ANSWERS

Exam-style question

Five girls each buy a gift. The gifts are collected together and they each choose one at random. The probability distribution for the number of girls, X, who end up with the gift they originally bought is shown in the table below.

r	0	1	2	3	4	5
$P(X=r)$	p	$\frac{3}{8}$	$\frac{1}{6}$	$\frac{1}{12}$	0	$\frac{1}{120}$

i   Find the value of p.
ii Explain why it is impossible for X to be 4.
iii Show that $P(X=5)=\frac{1}{120}$.

Short answers on page 159

Full worked solutions online

CHECKED ANSWERS

Chapter 4 The binomial distribution

About this topic

Many of the situations where you can use tree diagrams involve two possibilities at each stage. If the probabilities stay constant throughout the tree diagram, this is a binomial situation. The binomial probability model will be developed in this chapter to allow you to calculate such probabilities quickly.

Using the binomial probability formula enables you to work out the probability of one outcome. If you want several outcomes combined together, it is much quicker to use the probability functions on your calculator.

Before you start, remember ...

● Probability; tree diagrams; discrete random variables (all revised in Chapter 3).

Introducing the binomial distribution

Key facts

● In a binomial situation, these criteria apply:
  ○ You are conducting an experiment or trial n times (a fixed number) e.g. tossing a coin eight times.
  ○ There are two outcomes, which you can think of as 'success' and 'failure' e.g. heads and tails.
  ○ The probability of 'success' is the same each time (symbol p). The probability of 'failure' is $q = 1 - p$.
  ○ The probability of 'success' on any trial is independent of what has happened in previous trials.
  ○ The random variable, X, is 'the number of successes'.
● You write $X \sim B(n, p)$ to show that X has a binomial distribution. The values of n and p are called the **parameters** of the binomial distribution.
● $P(X = r) = {}_nC_r q^{n-r} p^r$ where $q = 1 - p$. $r = 0, 1, 2, ..., n$. ${}_nC_r = \dfrac{n!}{r!(n-r)!}$.

Worked example

Example 1

A biased coin has probability 0.4 of landing heads. It is tossed three times. What is the probability that it lands heads twice?

Solution

This can be solved by using a tree diagram.

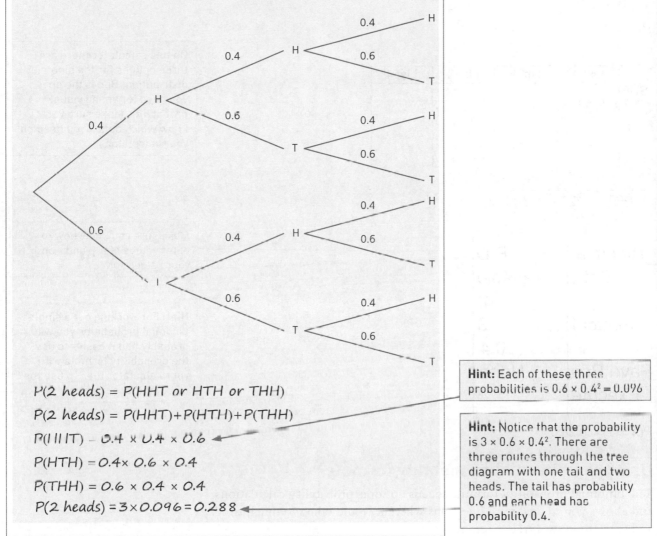

$P(2 \text{ heads}) = P(HHT \text{ or } HTH \text{ or } THH)$

$P(2 \text{ heads}) = P(HHT) + P(HTH) + P(THH)$

$P(HHT) = 0.4 \times 0.4 \times 0.6$

$P(HTH) = 0.4 \times 0.6 \times 0.4$

$P(THH) = 0.6 \times 0.4 \times 0.4$

$P(2 \text{ heads}) = 3 \times 0.096 = 0.288$

Hint: Each of these three probabilities is $0.6 \times 0.4^2 = 0.096$

Hint: Notice that the probability is $3 \times 0.6 \times 0.4^2$. There are three routes through the tree diagram with one tail and two heads. The tail has probability 0.6 and each head has probability 0.4.

For the random variable X = number of heads, using the binomial probability formula with $n = 3$ and $p = 0.4$ gives this probability in one calculation.

$$P(X = r) = {}_nC_r q^{n-r} p^r \quad \text{where} \quad q = 1 - p$$

$$P(X = 2) = {}_3C_2 \times 0.6^{3-2} \times 0.4^2$$

$$= 3 \times 0.6 \times 0.4^2$$

$$P(X = 2) = 0.288$$

$q = 1 - 0.4 = 0.6$

This part of the formula tells you how many relevant routes there would be through a tree diagram.

This part of the formula tells you what the probability for each relevant route would be if you used a tree diagram.

Hint: When using the binomial probability formula, start by writing down what X, n, p and q stand for. Write down the formula and substitute the correct numbers into it then use your calculator to find the answer.

Using a calculator

It is possible to get binomial probabilities from calculators by inputting values of n, p and r. If you use a calculator always write down the binomial distribution you are using and the probability you are calculating.

For Example 1, you should write the following.

X = no of heads.

$X \sim B(3, 0.4)$

$P(X = 2)$.

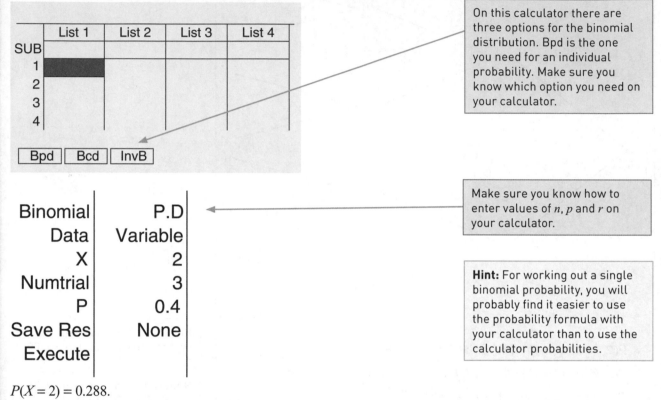

$P(X = 2) = 0.288$.

Using the binomial probability formula

The binomial probability formula speeds up some probability calculations and allows you to deal with questions where it would take too much time and space to draw a tree diagram.

Worked examples

Example 2

Adam throws ten dice and gets a score of 6 seven times. He wonders if there is something wrong with the dice. What is the probability of getting seven 6s when throwing ten fair dice?

Solution

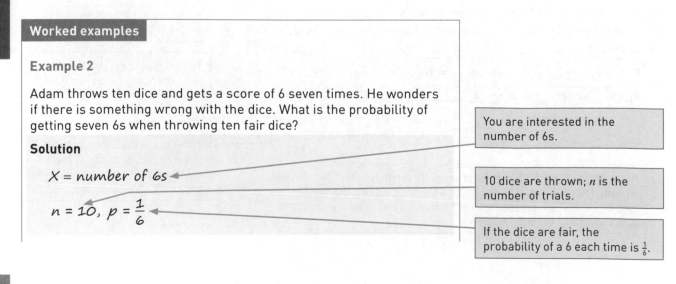

X = number of 6s

$n = 10$, $p = \dfrac{1}{6}$

Hint: Now that you have written down the probability distribution you are using with the correct values of n and p and the probability you are going to calculate, use your calculator to find the answer.

On this calculator there are three options for the binomial distribution. Bpd is the one you need for an individual probability. Make sure you know which option you need on your calculator.

Make sure you know how to enter values of n, p and r on your calculator.

Hint: For working out a single binomial probability, you will probably find it easier to use the probability formula with your calculator than to use the calculator probabilities.

You are interested in the number of 6s.

10 dice are thrown; n is the number of trials.

If the dice are fair, the probability of a 6 each time is $\frac{1}{6}$.

$$q = 1 - \frac{1}{6} = \frac{5}{6}$$

$$P(X = r) = {}_nC_r\, q^{n-r} p^r$$

$$P(X = 7) = {}_{10}C_7 \times \left(\frac{5}{6}\right)^3 \times \left(\frac{1}{6}\right)^7 = 0.000\ 248 \text{ (3 s.f.)}$$

This is the probability of not getting a 6.

You want the probability of seven 6s so $r = 7$.

Example 3

Adam thinks that one or two of the dice should land on 6. What is the probability of getting one or two 6s when throwing ten fair dice?

Solution

$$P(X = 1) = {}_{10}C_1 \times \left(\frac{5}{6}\right)^9 \times \left(\frac{1}{6}\right)^1 = 0.323\ 011 \ldots$$

$$P(X = 2) = {}_{10}C_2 \times \left(\frac{5}{6}\right)^8 \times \left(\frac{1}{6}\right)^2 = 0.290\ 710 \ldots$$

$$P(X = 1 \text{ or } 2) = 0.323\ 011 \ldots + 0.290\ 710 \ldots = 0.613\ 721 \ldots$$

$$P(X = 1 \text{ or } 2) = 0.614 \text{ (3 s.f.)}$$

Hint: Start by working out the probability of one 6 and two 6s separately.

Common mistake: Do not round before the final answer.

Finding the probability of 'at least one'

Hint: If you are asked to find the probability of at least one 'success', it is usually quickest to find the probability of no successes and subtract this from 1. There are either no successes or there is at least one, so these probabilities add to 1.

Worked example

Example 4

A multiple choice test has five questions, each with four possible answers. A student guesses all the answers. What is the probability he gets at least one correct?

Solution

$X =$ number of correct answers

$$n = 5, \quad p = \frac{1}{4}, \quad q = \frac{3}{4}$$

$$P(X = r) = {}_nC_r\, q^{n-r} p^r$$

$$P(X = 0) = {}_5C_0 \left(\frac{3}{4}\right)^5 \left(\frac{1}{4}\right)^0 = 0.237\ 304 \ldots$$

$$P(\text{at least one right}) = 1 - 0.237\ 304 \ldots = 0.762\ 695 \ldots$$

$$= 0.763 \text{ (3 s.f)}$$

Hint: Find the probability of getting none right first.

Common mistake: It is much easier to read an answer given to about 3 significant figures. Even though the probability with more decimal places is more accurate, it is helpful to round it as well as giving the full answer.

Recognising situations which are binomial

In an exam paper of mixed questions, it is not always obvious when you should use the binomial probability formula. It is even less obvious when using probability in real life. The key facts at the start of this section will help you to decide if it is a binomial situation.

Worked examples

Example 5

Frooties are a type of fruit flavoured sweet; they are sold in tubes of 14 sweets. Anya likes the red ones. 20% of Frooties produced are red. What is the distribution of the number of red sweets in a tube? What assumptions have you made in answering this question?

Solution

X = number of red sweets in a tube

$n = 14$
$p = 0.2$
$X \sim B(14, 0.2)$

Assumptions: The probability of each sweet being red is 0.2. This probability is independent of whether other sweets in the tube are red or not.

> n is the number of trials. There are 14 sweets in a tube and checking each one is a 'trial'.

> The proportion of red sweets is 0.2.

> This is how you write that X has a binomial distribution with $n = 14$ and $p = 0.2$.

> This may not be the case; it depends on how well the sweets are mixed before they are packed.

Example 6

Half the students in a class of 30 are girls. The teacher chooses students at random to answer questions but does not ask the same student twice in a lesson. He asks 12 questions in a lesson. Explain why you could not use a binomial distribution to find the probability that six girls and six boys are questioned.

Solution

The probability of the first student being a girl is $\dfrac{15}{30} = 0.5$

but, once the first student has been asked, the probability of the second student being a girl will be different because there are now only 29 students to choose from, and only 14 girls left. For a binomial distribution, the probability must be the same each time.

> **Common mistake:** Watch out very carefully to see if the probability of 'success' is the same each time, and independent of what has happened in previous trials.

Test yourself

1 A radio station is having a phone in programme. They guarantee that exactly 70% of calls, chosen at random, will be answered. Five friends ring up. What is the probability that exactly two of them have their calls answered?

   A 0   B 0.01323   C 0.1323

   D 0.3087   E 0.49

2 'One fifth of all chocolate bars contain a winning ticket' says the advert. Mark is determined to get a winning ticket. He buys ten chocolate bars. What is the probability that he has one or two winning tickets? (You should assume that the advert is true and that each bar of chocolate is equally likely to contain a winning ticket. Some answers are rounded to 4 d.p.).

   A 0.0336   B 0.0811   C 0.24

   D 0.5704   E 1

3 There are two tills in a shop. One of the assistants thinks that each customer is equally likely to go to either till. If she is right, what is the probability that, for the next ten customers, five go to each till?

A $\dfrac{1}{1024}$

B $\dfrac{63}{256}$

C $\dfrac{1}{32}$

D $\dfrac{1}{16}$

E 1

4 95% of the ball point pens produced in a factory work. They sell them in packs of six pens and give a money back guarantee if a customer buys a pack that contains any pens that do not work. What percentage of packs will they have to refund money on? (You should assume that the faulty pens are thoroughly mixed with the working pens before packing).

A 30%   B 26.5%   C 99.9%

D 23.2%   E 5%

5 The binomial distribution cannot be used for four of the random variables described below. It can be used for one of them. Find the one where the binomial distribution can be used.

A Five fair dice are thrown. The random variable is the total score on the dice.

B A bag contains a fixed number of balls. Some are black and the rest are white. A ball is taken without looking and its colour noted. It is replaced and the balls are mixed. This process takes place five times. The random variable is the number of times a white ball is seen.

C A fair coin is tossed until it lands heads. The random variable is the number of tosses.

D Students take a 40-question multiple choice test, for which they have revised. Each question has 5 possible answers. The random variable is the number of questions a student gets right.

E Rose opens a box of chocolates and eats them at random. There are 14 chocolates in the box, 4 are plain and the rest are milk. Rose eats 8 chocolates. The random variable is the number of plain chocolates she eats.

Full worked solutions online

CHECKED ANSWERS

Exam-style question

In the seventeenth century, people used to gamble on whether there would be at least one double 6 in 24 throws of a pair of dice.

i Find the probability of a double 6 when you throw a pair of fair dice once.

ii Find the probability of exactly one double 6 in 24 throws of the pair of dice.

iii Find the probability of at least one double 6 in 24 throws of the pair of dice.

Short answers on page 159

Full worked solutions online

CHECKED ANSWERS

Cumulative probabilities and expectation

REVISED

Key facts

- The expectation of a binomial random variable is $E(X) = np$.
- Make sure you know whether your calculator works out $P(X \leq x) = P(X = 0) + P(X = 1) + ... + P(X = x)$ or $P(a \leq X \leq b)$ and that you can use it to answer any question which involves the probability of a range of values of a binomial random variable.

The expectation of a binomial random variable

For a binomial discrete random variable, the expected value, or mean, is given by the following formula.

For $X \sim \text{B}(n, p)$, $\text{E}(X) = np$.

Example 1

A bank claims that 95% of callers to its helpline wait less than five minutes to speak to an adviser. A random sample of 40 callers is surveyed to find out how long it took before they were able to speak to an adviser. What is the expected number in the sample that waited at least five minutes?

> Remember to say what any letters you introduce stand for.

Solution

For a random sample, it is reasonable to assume that the probability for each caller of waiting longer than five minutes is the same and independent of other callers.

X = number of callers in the sample who waited at least five minutes

p = probability of a caller waiting at least 5 minutes = 0.05

$X \sim \text{B}(40, 0.05)$

$\text{E}(X) = np = 40 \times 0.05 = 2$.

> 95% less than five minutes means 5% at least five minutes.

> Show that you know this is a binomial distribution and give the parameters.

> **Common mistakes:**
> - If all callers one evening were surveyed instead of a random sample, it would not be reasonable to assume that the probability of each one waiting at least five minutes was the same. If one caller takes a long time on the phone, this will make the waiting time of the ones who come after him (or her) longer.
> - The expected value might not have a very high probability of occurring and it need not be a whole number.

Example 2

For the sample in Example 1, what is the probability that exactly two callers in this sample waited at least 5 minutes?

Solution

$X \sim \text{B}(40, 0.05)$

> Example 1 shows what the letters stand for, this is not repeated here.

$P(X = r) = {}_nC_r \, q^{n-r} \, p^r$ where $q = 0.95$

$P(X = 2) = {}_{40}C_2 \times 0.95^{38} \times 0.05^2 = 0.277\ 671...$

$\quad\quad\quad\quad = 0.278$ (to 3 s.f.).

> Putting the numbers into the binomial probability formula then using a calculator to get the answer.

Using your calculator to work out the total probability of a range of values for a binomial distribution

The binomial probability formula is quick to use if you just want a single probability. To save work when dealing with situations where several probabilities would need to be added, you should use the probability functions on your calculator. This is shown in the next two examples.

> **Hint:** Some calculators will work out $P(X \leq x) = P(X = 0) + P(X = 1) + ... + P(X = x)$ for a binomial distribution; these are usually scientific calculators. Other calculators will work out $P(a \leq X \leq b)$ for any range of values for a binomial distribution; these are usually graphical calculators. Make sure you know what your calculator does and can use it to answer any question which involves the probability of a range of values of a binomial random variable.

Worked examples

Example 3

An office has eight identical printers. On any given day, the probability of any one of them being faulty is 0.15, independently of the others. What is the probability that, on a given day, three or fewer of the printers are faulty?

Solution

X = the number of faulty printers on a given day.

$n = 8$, $p = 0.15$ where p is the probability of a printer being faulty.

X has a binomial distribution as the probability of each printer being faulty is the same so $X \sim B(8, 0.15)$.

You want $P(X \leqslant 3)$

Hints: Write what you are asked to find in symbols. Now use your calculator to get the answer.

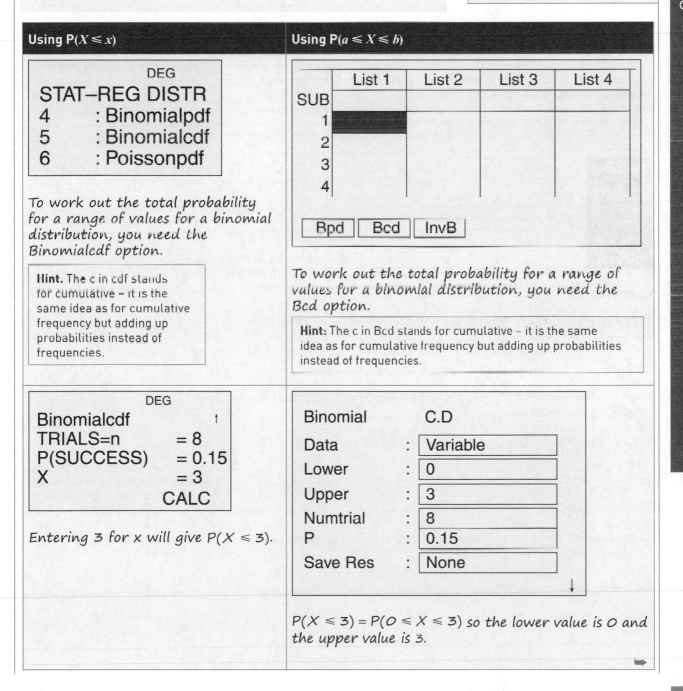

Using $P(X \leqslant x)$	Using $P(a \leqslant X \leqslant b)$

Using $P(X \leqslant x)$

```
                    DEG
STAT–REG DISTR
4       : Binomialpdf
5       : Binomialcdf
6       : Poissonpdf
```

To work out the total probability for a range of values for a binomial distribution, you need the Binomialcdf option.

Hint. The c in cdf stands for cumulative – it is the same idea as for cumulative frequency but adding up probabilities instead of frequencies.

```
                    DEG
Binomialcdf              ↑
TRIALS=n        = 8
P(SUCCESS)      = 0.15
X               = 3
                    CALC
```

Entering 3 for x will give $P(X \leqslant 3)$.

Using $P(a \leqslant X \leqslant b)$

	List 1	List 2	List 3	List 4
SUB				
1				
2				
3				
4				

```
Bpd    Bcd    InvB
```

To work out the total probability for a range of values for a binomial distribution, you need the Bcd option.

Hint: The c in Bcd stands for cumulative – it is the same idea as for cumulative frequency but adding up probabilities instead of frequencies.

```
Binomial        C.D
Data        :   Variable
Lower       :   0
Upper       :   3
Numtrial    :   8
P           :   0.15
Save Res    :   None
                            ↓
```

$P(X \leqslant 3) = P(0 \leqslant X \leqslant 3)$ so the lower value is 0 and the upper value is 3.

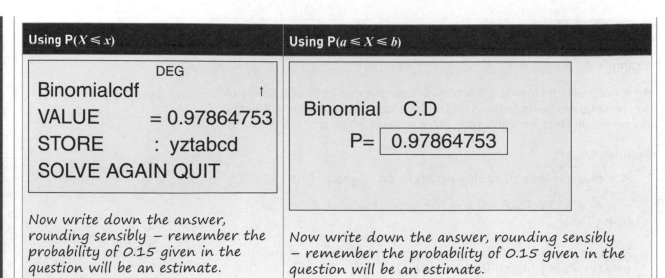

Using P($X \leqslant x$)	Using P($a \leqslant X \leqslant b$)

Binomialcdf DEG

VALUE = 0.97864753

STORE : yztabcd

SOLVE AGAIN QUIT

Binomial C.D

P= 0.97864753

Now write down the answer, rounding sensibly – remember the probability of 0.15 given in the question will be an estimate.

$P(X \leq 3) = 0.9786$

Now write down the answer, rounding sensibly – remember the probability of 0.15 given in the question will be an estimate.

Hint: You may find it helpful to write down the possible values that you are trying to find the probability of in order to help you to see how to enter the information into your calculator.

Common mistake: Be careful when writing the probability you want as an inequality. Some commonly used phrases, and how they translate to symbols, are listed in the table below.

Phrase	Less than 3	No more than 3	Up to 3	More than 3	No less than 3	At least 3
Symbols	$X < 3$	$X \leqslant 3$	$X \leqslant 3$	$X > 3$	$X \geqslant 3$	$X \geqslant 3$
Values (X is an integer)	0, 1, 2	0, 1, 2, 3	0, 1, 2, 3	4, 5, ...	3, 4, ...	3, 4, ...

Example 4

18 fair dice are thrown. What is the probability that there are more than six sixes?

Solution

X = the number of 6s; this has a binomial distribution as the dice are all fair.

$n = 18, p = \frac{1}{6}$, where p = probability of throwing a 6.

$X \sim B\left(18, \frac{1}{6}\right)$.

You want $P(X > 6)$. Possible values are 7, 8, 9, 10,..., 18. Hence, not 0, 1, 2,...,6.

Hint: Write down the possible values that you are trying to find the probability of in order to help you to see how to enter the information into your calculator.

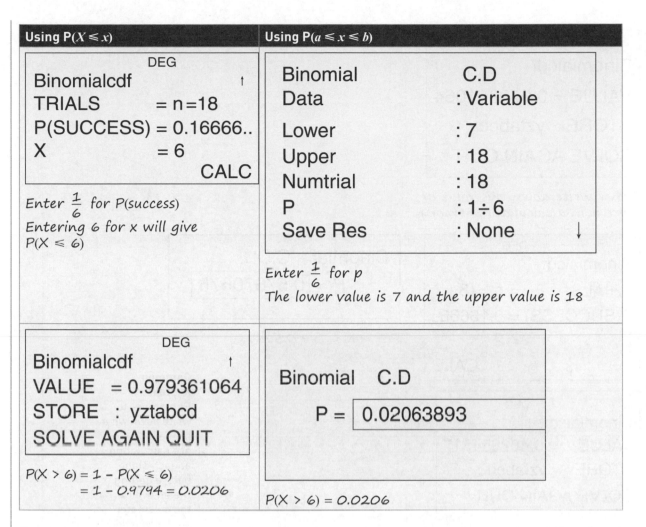

Using P($X \leqslant x$)

```
                    DEG
Binomialcdf              ↑
TRIALS        = n=18
P(SUCCESS) = 0.16666..
X             = 6
                    CALC
```

Enter $\frac{1}{6}$ for P(success)

Entering 6 for x will give P($X \leqslant 6$)

Using P($a \leqslant x \leqslant b$)

```
Binomial           C.D
Data             : Variable
Lower            : 7
Upper            : 18
Numtrial         : 18
P                : 1÷6
Save Res         : None        ↓
```

Enter $\frac{1}{6}$ for p

The lower value is 7 and the upper value is 18

```
                    DEG
Binomialcdf              ↑
VALUE  = 0.979361064
STORE   :  yztabcd
SOLVE AGAIN QUIT
```

P($X > 6$) = 1 – P($X \leqslant 6$)

= 1 – 0.9794 = 0.0206

```
Binomial    C.D

    P =  0.02063893
```

P($X > 6$) = 0.0206

Example 5

For the random variable, X, in Example 4, find P($3 \leqslant X < 7$).

Solution

P($3 \leqslant X < 7$)

Hint: The possible values are 3, 4, 5, 6

Using P($X \leqslant x$)

```
                    DEG
Binomialcdf              ↑
TRIALS        = n =18
P(SUCCESS) = 0.16666..
X             = 6
                    CALC
```

Enter $\frac{1}{6}$ for P(success)

Entering 6 for x will give P($X \leqslant 6$)

Using P($a \leqslant X \leqslant b$)

```
Binomial           C.D
Data             : Variable
Lower            : 3
Upper            : 6
Numtrial         : 18
P                : 0.16666666
 Save Res        : None ↓
 None   LIST
```

The lower value is 3 and the upper value is 6

Remember that lower and upper values are inclusive for this kind of calculator

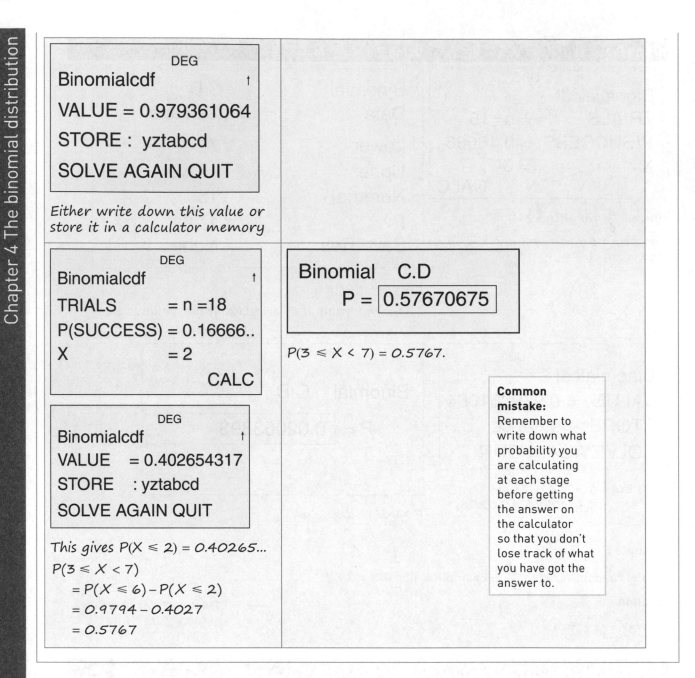

DEG

Binomialcdf ↑

VALUE = 0.979361064

STORE : yztabcd

SOLVE AGAIN QUIT

Either write down this value or store it in a calculator memory

DEG

Binomialcdf ↑

TRIALS = n = 18

P(SUCCESS) = 0.16666..

X = 2

CALC

Binomial C.D
P = 0.57670675

P(3 ≤ X < 7) = 0.5767.

DEG

Binomialcdf ↑

VALUE = 0.402654317

STORE : yztabcd

SOLVE AGAIN QUIT

This gives P(X ≤ 2) = 0.40265...
P(3 ≤ X < 7)
= P(X ≤ 6) − P(X ≤ 2)
= 0.9794 − 0.4027
= 0.5767

Common mistake: Remember to write down what probability you are calculating at each stage before getting the answer on the calculator so that you don't lose track of what you have got the answer to.

Test yourself

TESTED

1 In the game 'find the lady' the player has to say which one of three shuffled cards is the queen. The cards are face down. A player plays this game eight times. What is the expected number of times he wins, assuming that he is guessing?

A 2   B 2.5   C $2\frac{2}{3}$

D 3   E 4

2 X is a binomial random variable. Below are five pairs of inequalities; one in words and one in symbols. Four of the pairs are equivalent and one pair is not equivalent. Find the pair that is not equivalent.

A X is no more than 6; $X \leq 6$.

B X is at least 5; $X \geq 5$.

C X is more than 6; $X > 6$.

D X is not less than 7; $X \geq 7$.

E X is at most 7; $X > 7$.

3 X is a binomial random variable. Below are five statements about probabilities involving inequalities. Four of them are true and one is false. Find the one that is false.

A $P(X \geqslant 4) = 1 - P(X \leqslant 4)$

B $P(X < 3) = P(X \leqslant 2)$

C $P(X > 5) = 1 - P(X \leqslant 5)$

D $P(3 < X < 8) = P(X \leqslant 7) - P(X \leqslant 3)$

E $P(2 \leqslant X \leqslant 6) = P(X \leqslant 6) - P(X \leqslant 1)$

4 In the UK, 20% of children aged 2–15 years have asthma. Two adults are planning to take 12 children from this age group on a trip. Assuming that the children are a random sample from the population, what is the probability that no more than 2 out of the 12 children will suffer from asthma?

A 0.4896   B 0.2835   C 0.4417

D 0.5583   E 0.5584

5 There are ten hurdles in a race. A particular runner has a 65% chance of clearing any one of them. This probability is not affected by whether or not she has cleared the previous hurdles. What is the probability that she clears more than half the hurdles, but not all of them? (All answers are rounded to 3 d.p.)

A 0.500   B 0.892   C 0.738

D 0.741   E 0.905

Full worked solutions online

CHECKED ANSWERS

Exam-style question

A computer chooses 12 digits at random. Each digit can be either 0 or 1 and is equally likely to be either of these. The computer then adds all 12 digits to give a total.

i Explain why the total of these digits, X, is binomially distributed, $X \sim B(n,p)$, with $n=12$ and $p=\frac{1}{2}$.

ii What is the expected value of the total?

iii Find the probability that the total is at least 9.

Short answers on page 159

Full worked solutions online

CHECKED ANSWERS

Chapter 5 Statistical hypothesis testing using the binomial distribution

About this topic

A common and widely used statistical method is to take a sample and to use it to draw possible conclusions (inferences) about the population from which it is drawn. However, there is always a possibility that those conclusions, based on sample data, are wrong or inaccurate. A hypothesis test uses probability to assess whether a proposed model is consistent with the sample data. To take a simple example, if a coin lands heads 14 times and tails 6 times, is this consistent with the coin being fair or is there enough evidence to convince us that it is biased?

In a 1-tail test, you are looking for evidence in a particular direction, for example that the coin is biased towards heads. For a 2-tail test you are looking for evidence of any difference rather than for evidence of a difference in a particular direction.

Before you start, remember ...

- Binomial probability, including use of calculators (this is revised in Chapter 4).

Introducing hypothesis testing using the binomial distribution (1-tail tests)

Key facts

- In hypothesis testing for a binomial distribution, you use evidence from a sample to make a decision about the probability of 'success' for the whole population.
- For a hypothesis test using the binomial distribution, the null hypothesis is written in the form $H_0 : p = 0.3$ (or some other specific value).
- There should be a statement defining what p stands for, e.g. $p =$ the probability of a light bulb lasting less than 100 hours.
- For a 1-tail test the alternative hypothesis that goes with the example null hypothesis above is EITHER $H_1 : p > 0.3$ OR $H_1 : p < 0.3$. 1-tail tests are covered in this section.
- For a 2-tail test the alternative hypothesis that goes with the example null hypothesis above is $H_1 : p \neq 0.3$. 2-tail tests are covered in the next section.
- You can never be certain of the outcome from a hypothesis test but you can say that it is very unlikely that the null hypothesis was true. The **significance level** is the probability that you reject the null hypothesis even though it is true. It is often set at 5% or 1%.
- There are two main ways of making a decision in hypothesis testing.
  - o Using **probability**. The probability of the observed result or a more extreme result is compared with the significance level. This probability is called the **p-value**; if it is less than the significance level, H_0 is rejected, otherwise it is accepted.
  - o Using the **critical region**. The critical region is the set of values of X for which the probability of an extreme outcome is less than the significance level. If the observed value lies in the critical region, H_0 is rejected, otherwise it is accepted.
- The hypothesis test should end with a non-assertive conclusion in context.

You need to set out your working carefully when testing a hypothesis. This will help you to organise your thinking as well as getting you marks in the examination. The stages in the process are illustrated in the following situation.

The head of a large school announces that to give equal opportunities to all students, the School Council will be chosen at random from all the students, instead of being elected, because an election is biased towards popular students. There are equal numbers of boys and girls in the school but the School Council chosen consists of ten girls and four boys. Is there evidence (at the 5% level) of a bias towards girls in the selection process?

Deciding what the hypotheses are

If there is no bias, each vacancy is equally likely to be filled by a boy as it is to be filled by a girl. Let p = the probability of a vacancy being filled by a girl. The two possibilities are:

$p = \frac{1}{2}$ and $p > \frac{1}{2}$.

The first of these assumes the new policy is working so it is the null hypothesis. The hypotheses are:

$\left. \begin{array}{l} H_0 : p = \frac{1}{2} \\ H_1 : p > \frac{1}{2} \end{array} \right\}$ where p is the probability of a girl being chosen to fill a vacancy

> You are looking for evidence of a bias towards girls (and so evidence that the policy isn't working).

> **Hint:**
> - The null hypothesis is the belief you start with. You will only stop believing this if there is enough evidence.
> - The alternative hypothesis is what you are looking for evidence of.
> - For a hypothesis test using the binomial distribution, the null hypothesis must be: p = a specific value.
> - You must say what p stands for.
> - You will get marks in the exam for writing down the hypotheses correctly even if you go wrong with the rest of the question.

There are 14 vacancies. Assuming the null hypothesis is true, the number of girls on the council, X, would have a binomial distribution.

$X \sim B(14, 0.5)$ **if H_0 is true.**

> The null hypothesis, H_0, says that there is no bias.

Finding a critical region

To find the critical region, it is helpful to have a mental picture of the probability distribution.

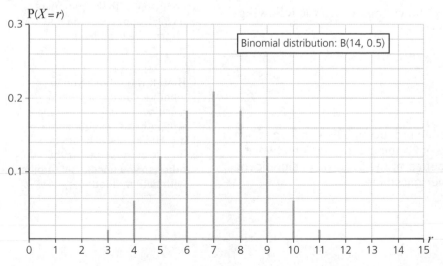

> **Hints:** You are looking for evidence of bias towards girls so need to find numbers of girls which are unusually large.

> **Hint:** The level at which you decide 'This is so unlikely that I do not believe the null hypothesis is true after all' is called the **significance level** of the test. In this case it was set at 5%, or 1 in 20. You are usually given the significance level in the question but sometimes you may be asked what it means as well.

Using a calculator, $P(X \geq 11) = 0.0286...$

Using a calculator, $P(X \geq 10) = 0.0897...$

Common mistake: Write down all the probabilities you work out when looking for a critical region. Even though you do not use the probability of 0.0897... it is important that you worked it out because this is what tells you that the critical region is $X \geq 11$.

The critical region is $X \geq 11$.

The number of girls on the School Council was 10. This is not in the critical region so you do not reject the null hypothesis. There is not enough evidence, at the 5% level, to show that there has been any bias towards girls.

Calculating a *p*-value

The increasing use of software in statistics has led to the use of *p*-values in hypothesis testing becoming more common.

The observed value was $X = 10$ and this raised a doubt as to whether H_0 is true; you would be even more doubtful if there were more girls so find $P(X \geq 10)$.

Using a calculator,

$P(X \geq 10) = 0.0898$.

The *p*-value is 0.0898; this is more than the significance level. There is not sufficient evidence of a bias towards girls at the 5% level of significance.

Common mistake: You can never be certain whether the null hypothesis was true so you should not make definite claims in your conclusion – state the conclusion in terms of there either being or not being enough evidence of what the alternative hypothesis claimed.

Hint: Compare the probability you work out with the significance level and reject H_0 if the probability is smaller than the significance level. This is equivalent to supposing that the critical region is $X \geq 10$ for Example 1, calculating the significance level and comparing it with the significance level that is required.

Putting it all together (using a *p*-value)

All the stages in a hypothesis test are shown together in the following example. This example uses a *p*-value.

Worked example

Example 1

A plastic moulding machine has been producing 30% misshapes. After the machine is serviced, a sample of 20 objects is taken and two of them are misshapes. Is there evidence, at the 5% level of significance, that the machine is producing fewer misshapes?

Hint: The low probability shows that getting 11 or more girls on the School Council would be unusual if there was no bias. 0.0286... is not 5% (the significance level of the test) so you should try looking at a larger 'tail'.

0.0897 is bigger than 5%. It is not possible to get exactly 5% so use the 'tail' with a lower probability than 5% rather than the one with a higher probability.

The critical value is 11; this is the boundary of the critical region.

Common mistake: Do not work out the probability of just one value, $P(X = 10)$ in the current example, because the probability for any particular value of X can be very small (especially if n is large) so you could be in the position of deciding that none of the outcomes are likely, but something has to happen. Always work out the *p*-value by finding probability of the observed outcome **together** with those which cast even more doubt on the null hypothesis.

Common mistake: The *p*-value is the probability of getting an outcome at least as extreme as that observed, if the null hypothesis is true. Don't confuse it with the parameter p which appears in the null and alternative hypotheses.

Hint: It will often be up to you whether you use a *p*-value or a critical region when doing a hypothesis test but some questions may ask for you to do one of these rather than the other so you need to be familiar with both methods.

Solution

Remember the null hypothesis starts $p = \ldots$.

$H_o : p = 0.3$ where p is the proportion of misshapes
$H_1 : p < 0.3$ produced by the machine

Remember to say what letters stand for.

X = number of misshapes in the sample of 20

You were asked to look for evidence that the machine is producing fewer misshapes.

If the null hypothesis is true, X has a binomial distribution with $n = 20$, $p = 0.3$.

$X \sim B(20, 0.3)$

Write down the distribution X would have if H_0 were true.

Need to find $P(X \leqslant 2)$.

Using a calculator,

$X = 2$ is what happened. A smaller value of X would make you believe the alternative hypothesis even more.

$P(X \leqslant 2) = 0.0355 = 3.55\%$. This is smaller than 5% so reject H_o.

There is sufficient evidence, at the 5% level of significance, to suggest that the machine is producing fewer misshapes.

State the conclusion non-assertively in the context of the original question.

Common mistakes:

- You know that 2 out of 20 in the sample were misshapes. That is 10% of the sample but you want to know about the percentage of misshapes in all the objects the machine produces. Is it likely to be less than 30%, or have you just got a good sample by chance?

- Statistical hypothesis testing does not allow you to be sure whether the machine is producing fewer misshapes or not; it provides evidence to help you make a decision.

A diagrammatic representation

For the situation in Example 1, the probabilities for the different values that X can take, if the null hypothesis is true, are shown in the vertical line chart:

Binomial distribution B(20, 0.3)

$P(X \leqslant 2)$ is shaded in purple. The total of these probabilities is less than 5%.

Hint: You may find a mental picture of the individual probabilities, helps you to see what is happening

Putting it all together (using a critical region)

Worked example

Example 2

An IT company knows its customers are dissatisfied with the helpline so it decides to train its staff. First the company surveys its customers and finds that 30% of them are dissatisfied with the helpline. Following training for the staff on the helpline, the company surveys a random sample of 19 customers and finds that 3 of them are dissatisfied with the helpline. It carries out a suitable hypothesis test.

i State the null and alternative hypotheses.

ii Find the critical region for the relevant test at the 5% level of significance.

iii Use the critical region to decide whether there is evidence that fewer of the customers are dissatisfied with the helpline.

Solution

i p = the proportion of customers who are dissatisfied with the helpline.

The hypotheses are:
$H_0 : p = 0.3$
$H_1 : p < 0.3$

> The company is looking for evidence that fewer customers are dissatisfied

X = the number of dissatisfied customers in a sample of size 19.

> Remember to say what letters stand for

ii If the null hypothesis is true, X will have a binomial distribution with $n = 19$ and $p = 0.3$.

This is written $X \sim B(19, 0.3)$.

> To find the critical region, it is helpful to have a mental picture of the probability distribution.

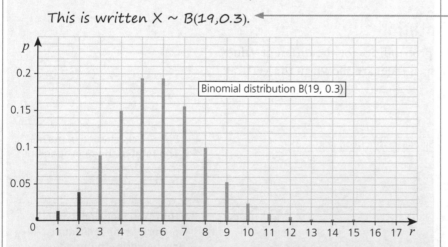

Binomial distribution B(19, 0.3)

The null hypothesis will be rejected if there is a small number of dissatisfied customers in the sample. Your task is to find low possible values of X with a total probability of 5%. This is sometimes called the lower tail.

Using a calculator,

$P(X \leqslant 2) = 0.0462$

$P(X \leqslant 3) = 0.1332$

> **Hint:** You need to write down both these probabilities, one each side of 5%, so that it is clear how you chose your critical region.

The first of these probabilities is 4.62%; the second is 13.32%.

It is not possible to get exactly 5% so go for 4.62%. The critical region is $X \leq 2$.

iii The number of dissatisfied customers in the sample of 19 was 3 so $X = 3$. This is not in the critical region so you accept the null hypothesis. There is not enough evidence, at the 5% level, to show that there has been a reduction in the number of dissatisfied customers.

Empty critical region

Worked example

Example 3

10% of the population of a country suffer from a particular chronic disease. Following a local public health campaign to reduce levels of the disease, a health worker wants to know whether the campaign has been effective. She decides to take a random sample of 20 people and test them for the disease to see if there is evidence, at the 5% level, that the proportion of people with the disease has dropped. Find the critical region.

Solution

The hypotheses are:

$H_o : p = 0.1$
$H_1 : p < 0.1$

where p is the proportion of people in the area with the disease
X = the number of people with the disease in a sample of 20

If the null hypothesis, H_o, is true, X has a binomial distribution with $n = 20$ and $p = 0.1$ which can be written $X \sim B(20, 0.1)$.

From a calculator,

$P(X = 0) = 0.1216 = 12.16\%$

This is more than 5% so the critical region is empty.

Common mistake: It is almost never possible to get exactly 5% (or whatever the significance level is) when finding a critical region for a binomial hypothesis test. You should always go for a probability lower than the significance level not the one nearest to it. In Example 2, the 10% significance level critical region would still be $X \leq 2$ because $X \leq 3$ has a probability of 13.32% and this is too high.

She is looking for evidence that the proportion has dropped below 10%.

Hint: The empty critical region means you cannot get enough evidence to reject H_0, at the 5% level, from a sample of 20 people. You need a larger sample for this test.

Test yourself

The following situation is for Questions 1 to 4. A local authority claims that two thirds of households in their area recycle glass. An environmentalist thinks that this is based on out of date data and that more households now recycle glass. He takes a random sample of 20 households and finds that 16 recycle glass. He wishes to test, at the 5% level of significance, whether there is evidence that more than two thirds of households in the area recycle glass.

1 Which of these gives the hypotheses for the test which the environmentalist uses?
In each case, p = proportion of households in the area that recycle glass.

A $H_0 : p > \frac{2}{3}$ B $H_0 : p = \frac{2}{3}$ C $H_0 : p = \frac{2}{3}$
$H_1 : p = \frac{2}{3}$ $H_1 : p > \frac{2}{3}$ $H_1 : p \neq \frac{2}{3}$

D $H_0 : p < \frac{2}{3}$ E $H_0 : p = 0.8$
$H_1 : p > \frac{2}{3}$ $H_1 : p < 0.8$

2 X = the number of households in the sample that recycle glass. Which one of the following probabilities will give the p-value for the hypothesis test?

A $P(X > 16)$ B $P(X = 16)$ C $P(X \geq 16)$

D $P(X < 16)$ E $P(X \leq 16)$

Before doing Question 3, you need to get the correct p-value.

3 Which is the decision the environmentalist should come to, with the correct reason?

A Accept H_0 because the p-value is more than 5%.

B Reject H_0 because the p-value is more than 5%.

C Accept H_0 because the p-value is different from 5%.

D Reject H_0 because the p-value is different from 5%.

E Reject H_0 because $\frac{16}{20} > \frac{2}{3}$.

4 Which of the following is the best statement of the final conclusion to the test?

A Accept H_0 at the 5% significance level.

B Reject H_0.

C It is certain that no more than two thirds of households recycle glass.

D There is insufficient evidence, at the 5% significance level, to justify the claim that more than two thirds of households recycle glass.

E Accept the alternative hypothesis at the 5% significance level.

5 A binomial hypothesis test is conducted with a sample size n, significance level 5% and hypotheses:
$H_0 : p = \frac{1}{3}$
X is the number of successes.
$H_1 : p < \frac{1}{3}$.

Four of the following statements about critical regions are true and one is false. Find the one that is false.

A There is only one value in the critical region for $n = 12$.

B If $n = 18$, the critical region is the same as it would be for a significance level of 10%.

C $n = 8$ is the smallest value of n for which the critical region is non-empty.

D For $n = 16$, the critical region is $X \leq 2$.

E The critical region is the same for each of $n = 8, 9, 10, 11, 12$.

Full worked solutions online

Exam-style question

A medical researcher claims that a new treatment is effective in 75% of cases. In an independent trial, a sample of 20 people try the treatment and it is effective for 12 of them. A suitable statistical test will be carried out to see whether there is evidence that the researcher's claims are exaggerated.

i   Write down suitable null and alternative hypotheses.
ii Carry out the test at the 5% level of significance, stating your conclusions clearly. You should assume that each person's response to the treatment is independent of the others.
iii Why is the assumption of independence important? Is it likely to be true?

Short answers on page 159

Full worked solutions online

CHECKED ANSWERS

2-tail tests

REVISED

Key facts

- For a hypothesis test using the binomial distribution, the null hypothesis is written in the form $H_0: p = 0.3$ (or some other specific value).
- For a 1-tail test the alternative hypothesis that goes with the example null hypothesis above is EITHER $H_1: p > 0.3$ OR $H_1: p < 0.3$. 1-tail tests were covered in the previous section.
- For a 2-tail test the alternative hypothesis that goes with the example null hypothesis above is $H_1: p \neq 0.3$ (this includes both possibilities: $p > 0.3$ and $p < 0.3$). 2-tail tests are covered in this section.
- 2-tail tests can be carried out using either probability or critical regions.
- The significance level for a 2-tail test is split into two halves: one half for each tail.

Symmetrical and asymmetrical binomial distributions

The binomial distribution is symmetrical if $p = 0.5$. It is skewed for other values of p. This is illustrated in the vertical line graphs below.

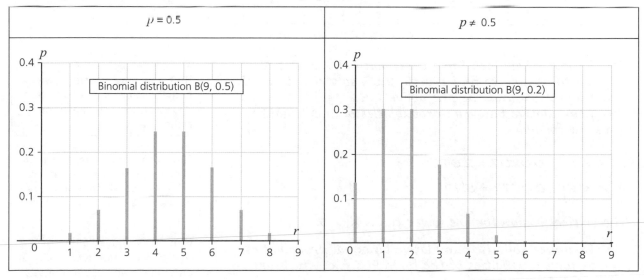

Critical region for a symmetrical 2-tail test

Worked example

Example 1

As part of an engineering project, Alicia invents a machine that is intended to replace tossing a coin at the start of a sports match. The machine should show 'heads' or 'tails' with equal probability. To test it for bias, she will run it 20 times. For what numbers of 'heads' will she conclude that it is biased, at the 5% level of significance?

Solution

$$H_o : p = 0.5$$

$$H_1 : p \neq 0.5$$

If the machine is unbiased, the probability of 'heads' on each go is 0.5.

She is looking for evidence of bias in either direction.

where p is the probability of 'heads' on each run.

If the null hypothesis is true and the probability of 'heads' is, indeed, 0.5 on each go then the number of heads in 20 runs will have a binomial distribution with $n = 20$, $p = 0.5$. This can be written as follows:

If H_o is true, $X \sim B(20, 0.5)$, where X is the number of 'heads' on 20 runs.

For a 2-tailed critical region at the 5% level of significance, you need 2.5% for each tail. The two tails are coloured in purple on the vertical line chart below.

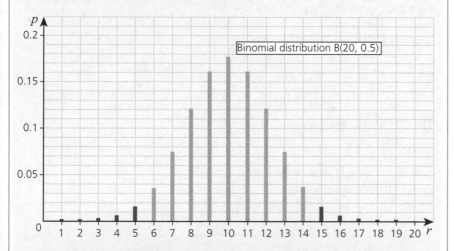

Binomial distribution B(20, 0.5)

For the lower tail, look for a probability of 2.5%; it is unusual to be able to find exactly 2.5% so write down the one below 2.5% and the one above 2.5%.
Using cumulative binomial probabilities on a calculator gives the following.

Hint: You need to write down both these probabilities so that it is clear how you chose your critical region.

$$P(X \leq 5) = 0.0207 = 2.07\%$$

$$P(X \leq 6) = 0.0577 = 5.77\%$$

Hint: Choose the one with a probability below 2.5% rather than the one with a probability above 2.5%.

The critical region for the lower tail is $X \leq 5$.

For the upper tail, use cumulative binomial probabilities to look for probabilities of the form $P(X \geq x)$ which are close to 2.5%.

Hint: You will need either to use $P(X \geq x) = 1 - P(X \leq (x-1))$ or enter lower and upper limits. Remember $n = 20$, $p = 0.5$.

$P(X \geqslant 16) = 0.0059$
$P(X \geqslant 15) = 0.0207$
$P(X \geqslant 14) = 0.0577$

The upper tail consists of the high values with a total probability of 2.5%, or less so the critical region for the upper tail is $X \geqslant 15$.

She should conclude the machine is biased if $X \leqslant 5$ or $X \geqslant 15$, where X is the number of 'heads' on 20 runs.

You need to write down both these probabilities so that it is clear how you chose your critical region.

Hints:
- If H_0 is $p = 0.5$, the critical regions for the upper and lower tails are symmetrical.
- For a 2-tailed critical region, work out half the significance level and then carry on as if you were working out two separate critical regions: one for $H_1 : p < 0.5$ and one for $H_0 : p > 0.5$.

Critical region for an asymmetrical 2-tail test

Worked example

Example 2

A seed firm sells a particular variety of flower seeds which have an 85% germination rate. The seeds are expensive so they are stored in carefully controlled conditions. A batch of the seeds has been stored in different conditions and a sample of 19 of this batch is planted to see if the rate of germination has been affected.

i   What numbers of germinating seeds would lead to the conclusion, at the 10% significance level, that there has been a change in the germination rate?

ii Hence, decide whether 18 seeds germinating would provide evidence of such a change.

Solution

i $H_o : p = 0.85$
   $H_1 : p \neq 0.85$
   where p is the proportion of the seeds in the batch that will germinate.
   If the null hypothesis is true and the proportion of seeds in the batch germinating is, indeed, 0.85 then the number of germinating seeds from a sample of 19 will have a binomial distribution with $n = 19$, $p = 0.85$.
   This can be written as follows:
   If H_0 is true, $X \sim B(19, 0.85)$, where X is the number of seeds in the sample of 19 that germinate.
   For a 2-tailed critical region at the 10% level of significance, you need 5% for each tail.
   Using cumulative binomial probabilities on a calculator (with $n = 19$, $p = 0.85$) gives the following.
   $P(X \leqslant 12) = 0.0163 = 1.63\%$
   $P(X \leqslant 13) = 0.0537 = 5.37\%$
   The critical region for the lower tail is $X \leqslant 12$.
   For the upper tail, use cumulative binomial probabilities to look for probabilities of the form $P(X \geqslant x)$ which are close to 5%.
   $P(X \geqslant 18) = 0.1985$
   $P(X \geqslant 19) = 0.0456$
   The critical region for the upper tail is $X \geqslant 19$. This simplifies to $X = 19$ as only 19 seeds were planted so no more than 19 can germinate.

Hint: Write down the probability below 5% and the one above 5%.

Hint: Choose the one with a probability below 5% rather than the one with a probability above 5%.

You will need either to use $P(X \geqslant x) = 1 - P(X \leqslant (x-1))$ or to enter lower and upper limits. Remember $n = 19$, $p = 0.85$.

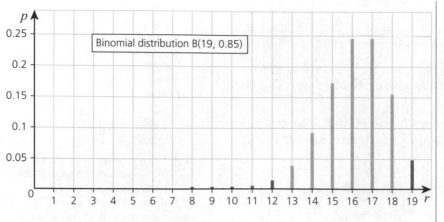

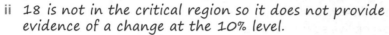

If 12 or fewer seeds germinate or if 19 seeds germinate, this will provide evidence, at the 10% level, that the germination rate has changed.

ii *18 is not in the critical region so it does not provide evidence of a change at the 10% level.*

Using probability to conduct a 2-tail hypothesis test

It is possible to conduct a 2-tail test by working out the probability of the 'tail'. It is not always obvious for an asymmetrical distribution whether a value lies in the upper or lower tail so work out the expected value first to help you decide.

The process is illustrated using the information from Example 2 above.

Expected value = np = 19×0.85 = 16.15

The observed value was 18; this is bigger than 16.15 so you want to look at the upper tail.

Now work out the tail probability that includes 18; this is
$P(X \geqslant 18) = 0.1985 = 19.85\%$

Since the test is a 2-tail test you should compare this with half the significance level.

19.85% > 5% so there is insufficient evidence of a change at the 10% level.

Test yourself

TESTED ☐

For questions 1 and 2, X is the number of successes for a binomial random variable and p is the probability of success on each trial.

1 A sample of size 19 is taken to test the hypotheses
$H_0 : p = 0.5$, $H_1 : p \neq 0.5$ at the 1% level of significance.

Which of the following gives the correct critical region?

A $X \leqslant 3, X \geqslant 15$ B $X \leqslant 3, X \geqslant 16$ C $X \leqslant 4, X \geqslant 15$

D $X \leqslant 5, X \geqslant 14$ E $X \leqslant 5, X \geqslant 16$

2 A sample of size 14 is taken to test the hypotheses
$H_0 : p = 0.6, H_1 : p \neq 0.6$ at the 10% level of significance.

Which of the following gives the correct critical region?

A $X \leqslant 4, X \geqslant 10$ B $X \leqslant 4, X \geqslant 11$ C $X \leqslant 4, X \geqslant 12$

D $X \leqslant 5, X \geqslant 12$ E $X \leqslant 3, X \geqslant 13$

The following information will be used in Questions 3 and 4:

As part of a psychology experiment, a random sample of 18 people was asked to choose one of two puzzles: either a normal sudoku, using numbers, or a puzzle which used letters instead of numbers, but which was otherwise identical to the first one. The experimenter wants to test whether there is a bias towards choosing either letters or numbers. The proportion of people in the population who would choose the puzzle with numbers is p. Thirteen people in the sample chose the puzzle with numbers. The test is conducted at the 5% level of significance.

3 Which of the following are the hypotheses for the test?

A $H_0 : p > 0.5$ B $H_0 : p > 0.5$ C $H_0 : p = 0.5$
    $H_1 : p < 0.5$ $H_1 : p = 0.5$ $H_1 : p > 0.5$

D $H_0 : p = 0.5$ E $H_0 : p \neq 0.5$
    $H_1 : p \neq 0.5$ $H_1 : p = 0.5$

Make sure you understand which are the correct hypotheses before attempting Question 4.

4 Which of the following is the correct conclusion for the test?

A   The critical region is $X \leq 4, X \geq 13$ so there is evidence of a bias.

B   The critical region is $X \leq 4, X \geq 14$ so there is not sufficient evidence of a bias.

C   The critical region is $X \leq 4, X \geq 14$ so there is evidence of a bias.

D   The critical region is $4 \leq X \leq 14$ so there is not sufficient evidence of a bias.

E   The critical region is $X \leq 5, X \geq 13$ so there is evidence of a bias.

5 A teacher reads an article which says that 15% of the population are left handed. She wonders whether the proportion is the same among students taking A Level Mathematics.
She observes 19 students taking a mathematics examination and notes how many of them are left handed (X). Assuming these 19 students are a random sample from the population of A Level Mathematics students, what is the critical region for the test at the 5% level of significance?

A $X \geq 4$ B   The critical region is empty

C $X \geq 6$ D $X = 0, X \geq 7$

E $X \geq 7$

Full worked solutions online

CHECKED ANSWERS

Exam-style question

A national survey shows that 15% of young people are trying to give up smoking. A local health authority wants to investigate whether the proportion of young people in its area who are trying to give up smoking is 15% or whether it is different from this. A random sample of 20 young people contains one who is trying to give up smoking.

i   Write down suitable hypotheses.

ii   What is the critical region for a suitable hypothesis test, at the 5% level?

iii   Carry out the test, stating your conclusions clearly.

Short answers on page 159

Full worked solutions online

CHECKED ANSWERS

Review questions (Statistics)

1 A researcher wants to know how many driving lessons learner drivers have before they pass their driving test. She chooses three driving schools from her local area and asks them to contact all their students who have passed a driving test in the last month.

  i Which of the following terms best describes the sampling method which the researcher is using?
opportunity, simple random, systematic, stratified, quota, cluster, self-selected

  ii The following table gives the data collected by the researcher. The drivers were asked to choose the group which contained the number of lessons they had.

Number of lessons	Frequency
0–9	0
10–30	5
31–40	8
41–50	16
51–60	14
61–100	10

Find an estimate of the mean number of lessons taken.

  iii It is suggested that the researcher should have asked for the exact number of lessons instead of asking the drivers to choose a group. Give one reason why the researcher might have asked the drivers to choose a group rather than giving the exact number of lessons.

2 The back to back stem-and-leaf diagram below shows the monthly average temperature in °C in Coventry for January and February near to the turns of the 20th and 21st centuries.

1896–1905		1996–2005	
3 1	1		
5 3	2	2 6	Scale
8 4 2 2	3	2 9	3| 2 | 2 represents 2.3°C for 1896–1905
7 5 3 3 2	4	1 2 2 7	and 2.2 °C for 1996–2005
5 5 2 0 0	5	1 1 1 4 4 5 8 8 9	
1	6	8	
1	7	0 3	

(Data from Bablake weather station)

  i Without doing any calculations, make two comments describing and comparing the two distributions.

  ii For 1896–1905 the median was 4.3°C and the interquartile range was 1.9°C.

Find the equivalent figures for 1996–2005.

  iii Decide whether the data provide evidence of global warming.

3 The trustees of a charity are going to appoint a full time manager. They decide to pay the manager the same as the average of the trustees' earnings. The annual incomes of the trustees are given in the list below (in pounds):

    44 000   39 600   17 600   77 000   26 400   110 000   90 000   31 080   28 600

Calculate:

i a The mean.

   b The median.

   c The mid-range.

ii An additional trustee is appointed; her income is lower than that of any of the other trustees. Explain how this additional data item affects each of the averages you calculated in part **i**.

iii Which of the three averages would be the most appropriate one to use when deciding the manager's income? Justify your choice.

4 Students at a large college have taken an English test and a mathematics test. 70% of students passed the English test; 80% of students passed the mathematics test. 15% of students did not pass either test.
A student is chosen at random from the college, what is the probability that the student has passed both the English test and the mathematics test?

5 Three dice are thrown. What is the probability that the highest score is 2?

6 Tanya has made a spinner which will show integers in the range 1 to 8 (inclusive). She thinks the spinner may be biased against the number 4 and wants to test if this is indeed the case. She spins it 30 times. Her results are shown on the dot plot below.

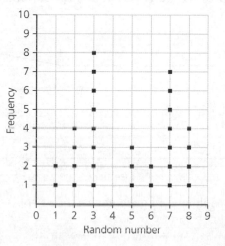

i Assume that each integer is equally likely. Find the probability of getting no fours in 30 spins.

ii Conduct a hypothesis test at the 5% significance level for the following hypotheses.

$$\left.\begin{array}{l} H_0 : p = \dfrac{1}{8} \\[2mm] H_1 : p < \dfrac{1}{8} \end{array}\right\} \text{where } p \text{ is the probability of a four.}$$

7 A market researcher wants to know what percentage of the adult population are vegetarian. He finds a website which says that 8% of the UK adult population are vegetarian. He thinks the percentage is different to this in his area of the country.

i State suitable hypotheses for a statistical hypothesis test.

The market researcher will interview a random sample of 100 adults from the area.

ii Find the critical region for the hypothesis test at the 5% level.

Short answers on page 159

Full worked solutions online

CHECKED ANSWERS

MECHANICS

1 Drawing a displacement–time graph
A child runs 100 m from A to B at a constant velocity of $4\,\text{m s}^{-1}$. She waits for 5 s at B and then runs back to A at $5\,\text{m s}^{-1}$.
i Find the total time the child takes from leaving A to returning there.
ii Draw a displacement–time graph for the child.

(see page 101)

2 Interpreting the gradient of a position–time graph
Hamish and his mother leave home together and go to a football field. Hamish runs on ahead and waits for his mother when he gets there. Their journeys are shown on the distance–time graph in the figure below.

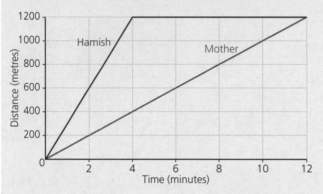

What is the difference in their speeds while they are travelling? Give your answer in m s^{-1} correct to 2 s.f.

(see page 103)

3 Interpreting the area under a velocity–time graph
Elizabeth takes 35.6 s to swim one length of a 50 m pool. The sketch graph in the figure below illustrates her motion; it is not drawn to scale but the line segments are indeed straight and the line that looks horizontal

is indeed horizontal. Elizabeth maintains a speed of $1.80\,\text{m s}^{-1}$ for T s. Find the value of T.

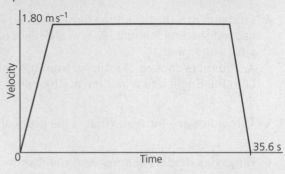

(see page 105)

4 Interpreting the gradient of a speed–time graph

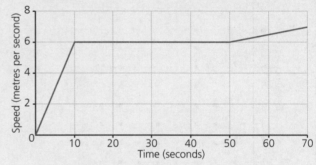

The graph in the figure above shows the speed of a runner in a race.
What was the magnitude of the runner's acceleration at each stage?

(see page 105)

5 Using constant acceleration formulae to find acceleration and distance travelled
A car is initially travelling at $10\,\text{m s}^{-1}$. The driver takes her foot off the accelerator and it comes to rest in 8 s. Assuming the acceleration of the car is constant, calculate:
i the acceleration
ii the distance the car travels while coming to rest.

(see page 111)

6 Using a constant acceleration formula to find time

A sledge is pulled with an acceleration of $1.5\,\mathrm{m\,s^2}$. The initial speed is $2\,\mathrm{m\,s^{-1}}$. Calculate the time taken to travel $20\,\mathrm{m}$.

(see page 112)

7 Finding the time taken and the final velocity for vertical motion under gravity

Freddie drops an egg onto the floor from a shelf $1.2\,\mathrm{m}$ above the ground. Find:
i   the time it takes to reach the ground
ii the velocity of the egg when it hits the ground.

(see page 115)

8 Finding initial velocity for vertical motion under gravity

Diego throws a ball into the air. It reaches a height of $8\,\mathrm{m}$ above the point of projection. Calculate the initial velocity of the ball.

(see page 116)

9 Using Newton's 3rd law

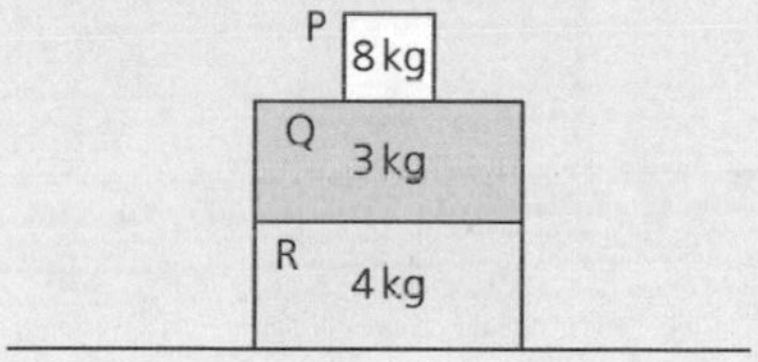

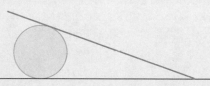

The diagram shows a pile of 3 blocks P, Q and R in equilibrium on a rough horizontal table. Their masses are $8\,\mathrm{kg}$, $3\,\mathrm{kg}$ and $4\,\mathrm{kg}$. The block R exerts a force N on the block Q. State the magnitude and direction of N.

(see page 122)

10 Identifying forces

A plank of wood rests on a fixed smooth cylinder with one end on rough horizontal ground as shown in the diagram. Show all the forces acting on the plank of wood.

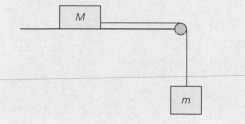

(see page 123)

11 Using Newton's 1st law for forces in equilibrium

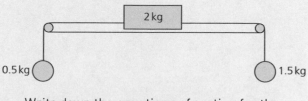

The diagram shows a block of mass $M\,\mathrm{kg}$ on a rough horizontal table. It is attached by a light string to a block of mass $m\,\mathrm{kg}$. The string passes over a light smooth pulley. The system is in equilibrium at rest.
Write down an expression for the frictional force acting on the block of mass on the table.

(see page 128)

12 Using Newton's 1st law for forces in two directions in equilibrium

In this question, $\mathbf{i}$ is a horizontal unit vector and $\mathbf{j}$ is the unit vector vertically upwards. A particle of mass $5\,\mathrm{kg}$ is in equilibrium under the action of three forces: its weight, W, and the tensions in two strings that are attached to it, $\mathbf{T_1}=3\mathbf{i}+y\mathbf{j}$ and $\mathbf{T_2}=x\mathbf{i}+15\mathbf{j}$.
Find the values of x and y.

(see page 128)

13 Using Newton's 2nd law to find acceleration

The mass of a spacecraft at the time it is launched is 50 tonnes. The propulsion unit produces an upwards force of $D\,\mathrm{N}$.
i   Find the greatest value of D for which lift-off will not be achieved.
ii The initial value of D is actually $800\,000\,\mathrm{N}$. Calculate the acceleration of the spacecraft.

(see page 130)

14 Using Newton's 2nd law to find a force

Lena pushes a toy train of mass $0.8\,\mathrm{kg}$ along a track with a horizontal force for $0.7\,\mathrm{s}$. The resistance to motion is $4\,\mathrm{N}$. The track is straight and horizontal and the train accelerates from an initial speed of $0.5\,\mathrm{m\,s^{-1}}$ and travels $98\,\mathrm{cm}$.
Calculate the pushing force that Lena applies.

(see page 131)

15 Connected particles

A block of mass $2\,\mathrm{kg}$ is placed on a smooth horizontal table and is attached to objects of mass $0.5\,\mathrm{kg}$ and $1.5\,\mathrm{kg}$ on opposite sides as shown in the diagram. The strings are light and inextensible and pass over smooth pulleys. The system is released from rest.

Write down the equations of motion for the block and each of the objects.

(see page 135)

16 **Using differentiation to determine acceleration**

The displacement of a particle at time t s is given by $s = 0.02t^3 - 0.8t^2 + 10t$ m. Determine whether the acceleration is constant.

(see page 140)

17 **Using differentiation to investigate motion**

Max runs in a straight line and his displacement in metres from the origin O at time t is given by $s = 10 + 0.72t^2 - 0.03t^3$ for $0 \le t \le 20$ where t is the time in seconds after the start of the run.

Determine whether Max changes direction during his 20 s run.

(see page 141)

18 **Motion with variable acceleration: integration**

A particle travels in a straight line and at time t s its acceleration in m s^{-2} is given by $a = 2t - 5$. When $t = 0$, the particle is at the origin with a velocity of 4 m s^{-1} in the positive direction. Find the distance of the particle from the origin when $t = 6$.

(see page 142)

Short answers on page 160

Full worked solutions online

CHECKED ANSWERS

Chapter 6 Kinematics

About this topic

Kinematics is the study of motion and so is an essential aspect of Mechanics. The language involved often requires the use of technical terms which need to be used precisely, for example the difference between vector and scalar quantities. It is common to supplement descriptions in words with graphs illustrating motion in particular cases and it is important to be able to draw and interpret such graphs.

In real life, motion is often complicated and it is necessary to use a simplifying model as a first step in analysing it. A very common model involves treating an object as a particle moving along a straight line with constant acceleration. This model can also be used for a large object when its dimensions and any bends in its path are negligible in terms of the overall motion.

You are forever seeing things falling, often after being thrown up in the air. This common situation can be well modelled as motion in a straight vertical line with a constant acceleration of g downwards due to gravity.

Before you start, remember ...

- The terms **time**, **distance** and **speed** (GCSE level)
- Knowledge of graphical techniques (GCSE level)
- Basic algebra (GCSE level)
- The gradient of a line
- The formula for calculating the area of a trapezium
- Simultaneous and quadratic equations
- The meanings of **position**, **displacement**, **distance**, **velocity**, **speed** and **acceleration**.
- How to solve equations including simultaneous and quadratic equations.
- Equations of motion in a straight line with constant acceleration (*suvat* equations)

Motion

REVISED

Key facts

VECTORS (have magnitude and direction)	SCALARS (have magnitude only)
Displacement	Distance
Position – displacement from a fixed origin	
Velocity – rate of change of position	Speed – magnitude of velocity
Acceleration – rate of change of velocity	
	Time

Warning: although acceleration is strictly a vector, it is commonly used as a scalar. The correct scalar term is magnitude of acceleration.

Definitions

- Average speed $=\dfrac{\text{total distance travelled}}{\text{total time taken}}$ (it is a scalar quantity)

- Average velocity $=\dfrac{\text{displacement}}{\text{time taken}}$ (it is a vector quantity)

- Average acceleration $=\dfrac{\text{change in velocity}}{\text{time}}$ (it is a vector quantity)

Graphs

- Position–time

- Velocity–time

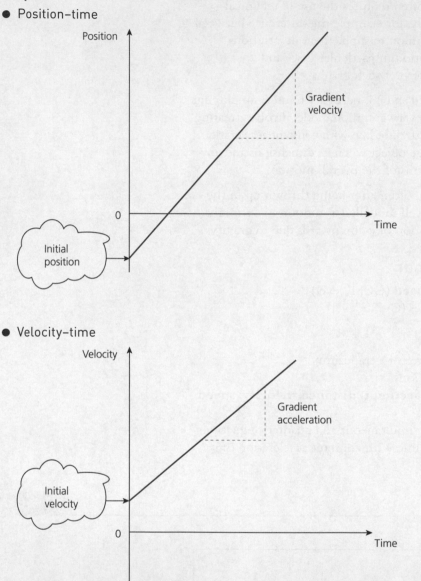

Worked examples

Example 1

The motion of a particle is illustrated by the position–time graph below.

  i Describe what is happening during these six seconds.

  ii Sketch a graph of distance travelled against time.

Solution

i

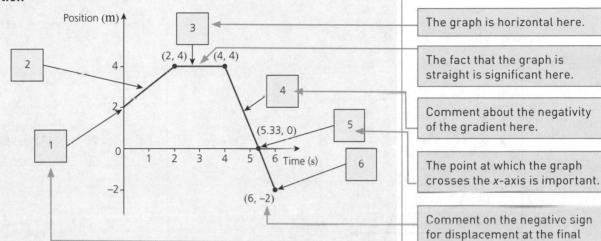

The graph is horizontal here.

The fact that the graph is straight is significant here.

Comment about the negativity of the gradient here.

The point at which the graph crosses the x-axis is important.

Comment on the negative sign for displacement at the final point.

Give the information about the initial point, where the graph cuts the y-axis

The five key elements of the motion during these six seconds are:

1 The particle starts 2 m from the origin;

2 The particle moves away from the origin with a constant speed ('constant' as the line is straight) of 1 m s^{-2} (the gradient of the line is $\frac{2}{2} = 1$ m s^{-1}) for 2 s; it travels 2 m;

3 For the next 2 s the particle remains at 4 m from the origin, so it is stationary;

4 The particle moves in the opposite direction with a constant speed of 3 m s^{-1} for 2 s, so the velocity is −3 m s^{-1} (the gradient of the line is −3);

5 The particle returns to its initial position after 5.33 s.

6 The particle ends up 2 m the other side of the origin.

ii The first section is a line is from (0, 0) to (2, 2) followed by a line from (2, 2) to (4, 2) and finally a line from (4, 2) to (6, 8)

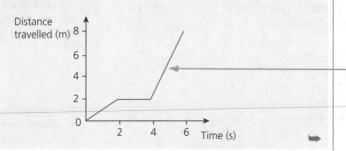

The key point here is that distance does not take into account the direction of travel, and so increases even when the particle is returning towards the origin.

Example 2

The position of a particle moving along a straight line is given by $x = 7 + t(6-t)$ ($0 \leqslant t \leqslant 7$, where x is measured in metres and t in seconds.

  i What is the position of the particle at times $t = 0, 1, 2, 3, 4, 5, 6$ and 7?

  ii Draw a diagram to show the position of the particle for $0 \leqslant t \leqslant 7$.

  iii Calculate the total distance travelled during the motion.

Solution

  i t: 0 1 2 3 4 5 6 7

     x: 7 12 15 16 15 12 7 0

> Substitute the values of t into the formula for x.

  ii

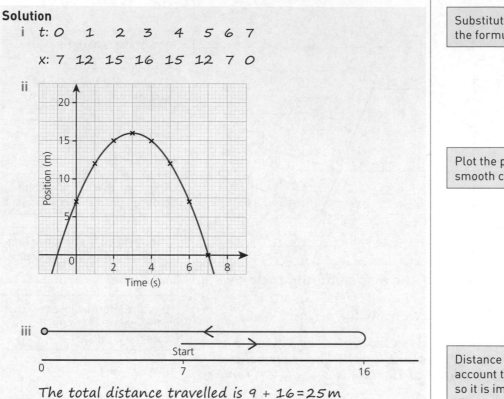

> Plot the points and join with a smooth curve.

  iii

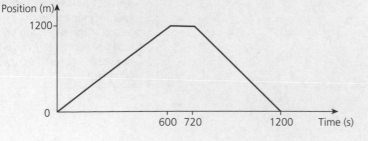

The total distance travelled is 9 + 16 = 25 m

> Distance does not take into account the direction of travel, so it is important to add the distances for the two phases of motion.

Example 3

Jack takes 10 minutes to walk 1200 m, at a constant rate, in a westerly direction. He rests for 2 minutes before taking another 8 minutes to return to his starting point, again walking at a constant rate.

  i Sketch a graph of position against time.

  ii Calculate his average speed for the whole journey.

  iii Calculate his average velocity for the whole journey.

Solution

  i Take the origin as his starting point, west as the positive direction and the unit of time as a second. As Jack walks at a constant rate, each section of the graph will be a straight line.

> Be clear where the origin is and which you are taking as the positive direction as Jack changes the direction of travel.

Position (m)

1200

0   600 720   1200   Time (s)

> When Jack is travelling back towards the origin, the gradient of the graph is negative.

ii Use average speed = $\dfrac{\text{total distance travelled}}{\text{total time taken}}$

So his average speed = $\dfrac{1200+1200}{600+120+480} = \dfrac{2400}{1200} = 2\,\text{ms}^{-1}$.

> Speed and distance are scalar quantities. The distances for each phase are added together.

iii Use average velocity = $\dfrac{\text{displacement}}{\text{time taken}}$

The displacement for the whole journey is $0\,\text{m}$ (west).

So his average velocity = $\dfrac{0}{20} = 0\,\text{ms}^{-1}$ (west).

> Velocity and displacement are vector quantities. The change in the direction of travel means that the overall displacement is zero.

Example 4

A train starts from rest and travels to the next station along a straight track. It accelerates uniformly at $0.2\,\text{ms}^{-2}$ for $150\,\text{s}$, then travels at a constant speed for the next $600\,\text{s}$ before slowing down at a constant rate, coming to a stop $100\,\text{s}$ later. What is the acceleration in the final stage of the journey?

Solution

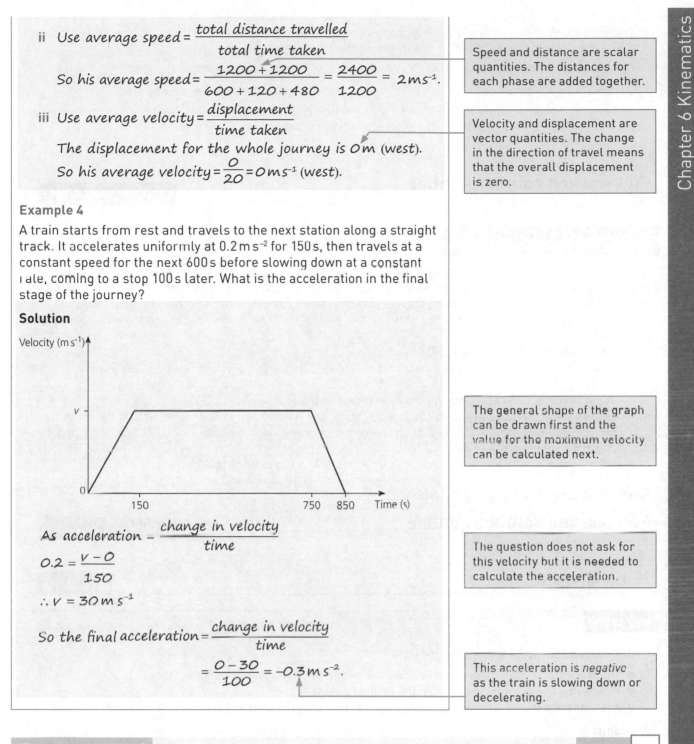

> The general shape of the graph can be drawn first and the value for the maximum velocity can be calculated next.

As acceleration = $\dfrac{\text{change in velocity}}{\text{time}}$

$0.2 = \dfrac{v-0}{150}$

$\therefore v = 30\,\text{m s}^{-1}$

> The question does not ask for this velocity but it is needed to calculate the acceleration.

So the final acceleration = $\dfrac{\text{change in velocity}}{\text{time}}$

$= \dfrac{0-30}{100} = -0.3\,\text{m s}^{-2}$.

> This acceleration is *negative* as the train is slowing down or decelerating.

Test yourself

TESTED ☐

1 Which one of the following statements is true?

A The speed of light is $3\times10^{8}\,\text{m s}^{-1}$ and the mean distance from the sun to the earth is $1.5\times10^{8}\,\text{km}$, so it takes $0.5\,\text{s}$ for light to reach the earth from the sun.

B The speed of sound is $340\,\text{m s}^{-1}$ and it takes $5\,\text{s}$ for the sound of thunder to reach me, so I must be $1.7\,\text{km}$ away.

C A particle has negative acceleration, so its velocity must also be negative.

D If a particle has zero velocity then its acceleration is also zero.

2 Alan walks $300\,\text{m}$ due east in $150\,\text{s}$, and then $150\,\text{m}$ due west in $50\,\text{s}$. What is his average velocity?

  A $2.25\,\text{m s}^{-1}$ B $0.75\,\text{m s}^{-1}$ C $2.25\,\text{m s}^{-1}$ east   D $0.75\,\text{m s}^{-1}$ east

3 The quantities in one of the following groups are either all scalars or all vectors. In the other groups there are some of each. In which group are all the quantities the same type?

A distance, velocity, acceleration

B time, displacement, speed

C time, speed, distance

D position, speed, acceleration

4 The position of a particle moving along a straight line is given by $x = 5 + 4t - t^2$, where x is measured in metres and t in seconds. What is the distance travelled between $t = 0$ and $t = 5$?

A −5 m   B 5 m   C 9 m   D 13 m

Full worked solutions online

CHECKED ANSWERS

Exam-style question

A shuttle train travels along a straight track between two stations A and B, which are 2.1 km apart. On leaving one station the train accelerates uniformly at $0.1\,\mathrm{m\,s^{-2}}$ and covers 500 m in 100 s before reaching its maximum speed of $v\,\mathrm{m\,s^{-1}}$. It then travels at a constant speed before beginning to slow down uniformly in order to stop at the other station. This final phase takes 80 s, during which time the train covers 400 m.

i a Calculate the maximum speed, $v\,\mathrm{m\,s^{-1}}$.

b How long does it take for the journey from A to B? What is the average speed of the journey from A to B?

c What is the acceleration during the final phase of the journey?

ii a Draw a velocity–time graph of the journey from A to B and back to A, assuming that the train stops for 3 minutes at station B and that on the return journey the pattern of motion is the same as when it is travelling from A to B.

b What is the average velocity for the journey from A to B and back to A?

Short answers on page 160

Full worked solutions online

CHECKED ANSWERS

Using graphs

REVISED

Key facts

1 **Definitions (for motion in a straight line)**
- **Displacement** – the distance and direction of one point from another (a vector)
- **Position** – the displacement from the origin (a vector)
- **Distance travelled** – the length of the path travelled, whatever the direction (a scalar)

2 **Graphs**

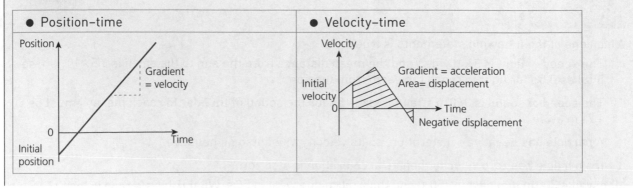

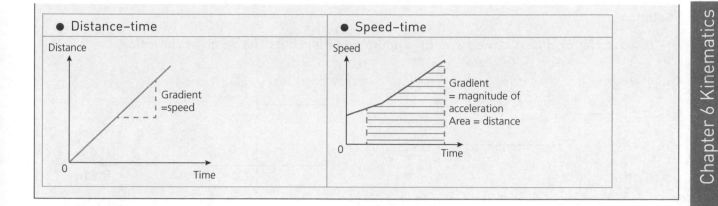

| ● Distance–time | ● Speed–time |

Worked examples

Example 1

A heavy goods train starts from rest and accelerates uniformly for 120 s, by which time its speed is 12 m s⁻¹. It travels at this speed for the next 600 s and then decelerates uniformly, coming to rest 80 s later.

i   Sketch the speed–time graph.

ii What is the acceleration of the train during the final stage of its journey?

iii Find the total distance travelled.

Solution

i

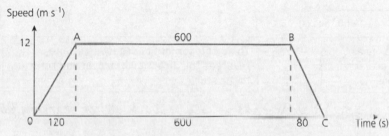

ii The acceleration during the final stage of the journey = gradient of BC

$$= \frac{-12}{80} = -0.15 \, m \, s^{-2}$$

The distance travelled is found by calculating the area under the speed–time graph, that is the area of the trapezium OABC.

Area = ½(800 + 600) × 12 = 8400

The distance travelled is 8400 m = 8.4 km.

> Uniform acceleration means that the graph is made from straight lines.

> **Common mistakes:** Note that the acceleration is *negative* as the train is slowing down.

> This area can also be found be splitting the area into two triangles and a rectangle.

Example 2

Jane walks North for 50 s at 3 m s⁻¹ and then South for 100 s at 2 m s⁻¹. Sketch:

i   the speed–time graph

ii the velocity–time graph

iii the distance–time graph

iv the position–time graph.

Solution

Choose the origin as Jane's starting point and North as the positive direction.

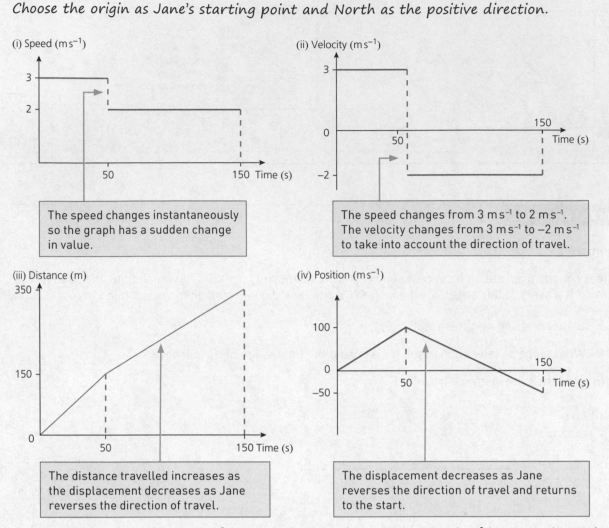

(i) Speed (m s⁻¹)

The speed changes instantaneously so the graph has a sudden change in value.

(ii) Velocity (m s⁻¹)

The speed changes from 3 m s⁻¹ to 2 m s⁻¹. The velocity changes from 3 m s⁻¹ to −2 m s⁻¹ to take into account the direction of travel.

(iii) Distance (m)

The distance travelled increases as the displacement decreases as Jane reverses the direction of travel.

(iv) Position (m s⁻¹)

The displacement decreases as Jane reverses the direction of travel and returns to the start.

Jane covers a total distance of 350 m. Jane ends up 50 m south of her starting point.

Example 3

A train takes 5 minutes to travel between two stations which are 2 km apart. The train accelerates for 60 s before reaching its maximum speed. It then travels at this speed before being brought to rest in 40 s with a constant deceleration. What is the maximum speed of the train?

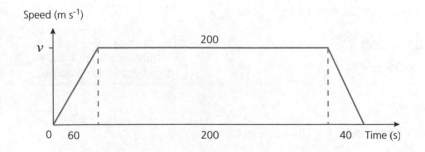

Speed (m s⁻¹)

Solution

The sketch of the speed–time graph of the journey shows the given information, with suitable units.

The maximum speed is v m s⁻¹.

The area is $\frac{1}{2}(300 + 200) \times v = 2000$

$$\Rightarrow v = \frac{2000}{250} = 8$$

> Take care with units here. The question has a mixture of minutes and seconds and metres and kilometres.

The maximum speed of the train is $8\,\text{ms}^{-1}$.

Example 4

The velocity–time graph illustrates the progress of a car along a straight road during a two minute period.

  i   Describe the car's journey

  ii Draw the acceleration–time graph.

  iii How far does the car travel?

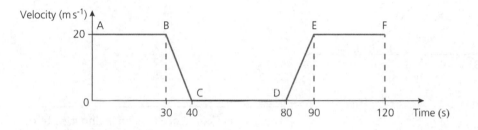

Solution

  i   AB: the car travels at a steady velocity of $20\,\text{ms}^{-1}$ for 30s

     BC: the car decelerates uniformly to a stop in 10s

     CD: the car is stationary for 40s

     DE: the car accelerates uniformly for 10s reaching a velocity of $20\,\text{ms}^{-1}$

     EF: the car travels at a steady velocity of $20\,\text{ms}^{-1}$ for 30s.

> Give a description for each phase of the journey.

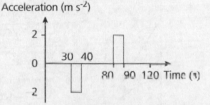

  ii In AB, CD and EF the acceleration is $0\,\text{ms}^{-2}$

     In BC the acceleration is $\frac{-20}{10} = -2\,\text{ms}^{-2}$

> Take care with signs – the acceleration is negative here.

     In DE the acceleration is $\frac{20}{10} = 2\,\text{ms}^{-2}$

> For each phase, the acceleration is constant. It changes from one value to the other instantaneously. In each case it is the gradient of the velocity–time graph.

  iii The displacement of the car is the same as the actual distance it travels as the velocity is not negative at any stage during the two minutes. It is found by calculating the area under the velocity–time graph.

     The velocity–time graph is symmetrical, so the area $= \frac{1}{2}(30 + 40) \times 20 \times 2$

$$= 1400$$

     The distance travelled is 1400m.

Example 5

The graph shows the motion of a particle. Use the graph to estimate the acceleration of the particle at $t=4\,$s.

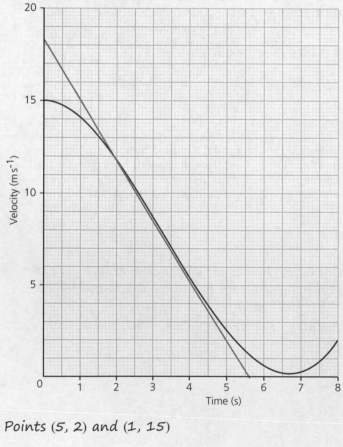

Solution

Draw the tangent to the curve at $t = 4$.

> The gradient of the graph is changing. Use the tangent to show the gradient at the required point.

Choose two points on the tangent that are easy to ready accurately and use them to calculate the gradient of the tangent.

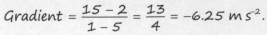

Points (5, 2) and (1, 15)

Gradient $= \dfrac{15 - 2}{1 - 5} = \dfrac{13}{4} = -6.25\ \mathrm{m\,s^{-2}}$.

> **Common mistake:** Take care with the signs here – check by looking at the graph that the gradient is supposed to be negative.

Test yourself

1 The driver of car, which has been stopped at traffic lights, finds that she is then stopped at the next traffic lights. Which of the following graphs could represent the velocity–time graph for the car as it travels between the two sets of traffic lights?

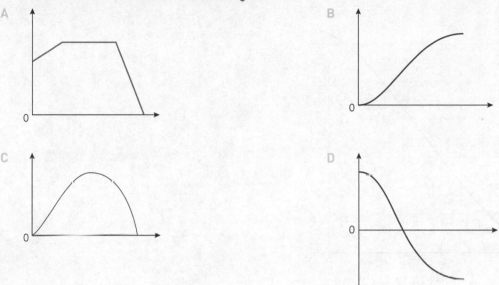

2 A particle moves in a straight line from A to C, passing through B. It starts from rest at A and accelerates uniformly at $1\,\mathrm{m\,s^{-2}}$ for 5s before arriving at B. Between B and C it accelerates uniformly at $2\,\mathrm{m\,s^{-2}}$ for another 5s. Draw the speed–time graph for this motion and then answer the following question. What is the average speed of the particle in its journey from A to C?

A $6\frac{2}{3}\,\mathrm{m\,s^{-1}}$ B $3.75\,\mathrm{m\,s^{-1}}$ C $1.5\,\mathrm{m\,s^{-1}}$ D $6.25\,\mathrm{m\,s^{-1}}$

3 A body travels in a straight line. It accelerates uniformly from rest at $0.4\,\mathrm{m\,s^{-2}}$, then travels at a constant speed of $8\,\mathrm{m\,s^{-1}}$, after which it is brought to rest with a constant retardation. It travels 600 m in 100s. Which of the following statements is true?

A The magnitude of the retardation is $0.4\,\mathrm{m\,s^{-2}}$.

B The body travels at a constant speed for 50s.

C The body travels half the distance in half the time.

D If the acceleration had been $0.8\,\mathrm{m\,s^{-2}}$ instead of $0.4\,\mathrm{m\,s^{-2}}$ the body would have travelled for longer at a constant speed.

4 A particle, which starts from rest, travels along a straight line. Its *acceleration* during the time interval $0 \leqslant t \leqslant 10$ is given by this acceleration–time graph. When is the speed of the particle greatest?

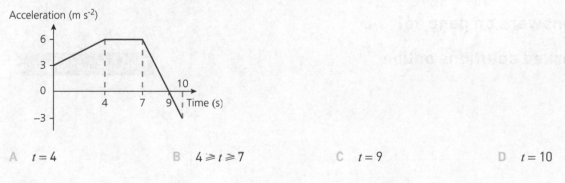

A $t = 4$ B $4 \geqslant t \geqslant 7$ C $t = 9$ D $t = 10$

5 The graph shows the displacement of a particle at time t. Use the graph to estimate the velocity of the particle at time $t = 4$ s.

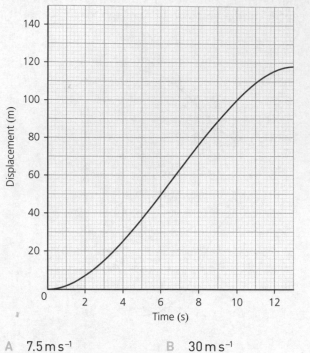

A   7.5 m s⁻¹   B   30 m s⁻¹   C   11.4 m s⁻¹   D   9.2 m s⁻¹

Full worked solutions online

CHECKED ANSWERS

Exam-style question

P and Q are two points 700 m apart on a straight road. A car passes the point P with a speed of 8 m s⁻¹ and immediately accelerates uniformly, reaching a top speed of 28 m s⁻¹ in 5 s. The driver continues at this speed for another 15 s before decelerating uniformly for T seconds at 1.8 m s⁻² until he reaches Q. When the car passes Q its speed is V m s⁻¹.

i   Sketch the speed–time graph for the journey between P and Q.
ii   Find:
   a   the acceleration
   b   the distance travelled from P to reach the top speed.
iii Find the values of T and V.
iv   a   Find the average speed for the journey between P and Q.
   b   Draw the acceleration–time graph for the journey from P to Q.

Short answers on page 161

Full worked solutions online

CHECKED ANSWERS

Using the constant acceleration formulae

Key facts

- The *suvat* equations
  The equations for motion with constant acceleration are

  $$s = \frac{1}{2}(u+v)t$$

  $$v = u + at$$

  $$s = ut + \frac{1}{2}at^2$$

  $$v^2 = u^2 + 2as$$

  $$s = vt - \frac{1}{2}at^2$$

  a is the constant acceleration;
  s is the displacement from the starting position at time t;
  v is the velocity at time t;
  u is the initial velocity (when $t = 0$).
  a and u have the same value throughout the motion. s and v vary as t varies.
  The *suvat* equations are obtained by assuming that the velocity–time graph is a straight line:
  The gradient is the acceleration and this is constant.

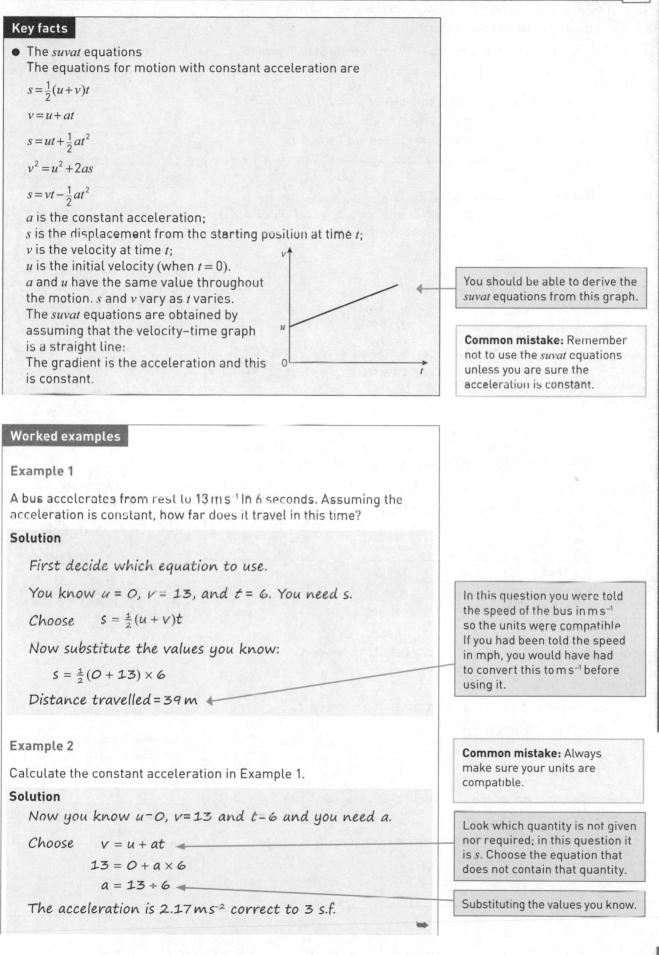

> You should be able to derive the *suvat* equations from this graph.

> **Common mistake:** Remember not to use the *suvat* equations unless you are sure the acceleration is constant.

Worked examples

Example 1

A bus accelerates from rest to $13\,\mathrm{m\,s^{-1}}$ in 6 seconds. Assuming the acceleration is constant, how far does it travel in this time?

Solution

First decide which equation to use.

You know $u = 0$, $v = 13$, and $t = 6$. You need s.

Choose $\quad s = \frac{1}{2}(u + v)t$

Now substitute the values you know:

$$s = \frac{1}{2}(0 + 13) \times 6$$

Distance travelled $= 39\,\mathrm{m}$

> In this question you were told the speed of the bus in $\mathrm{m\,s^{-1}}$ so the units were compatible. If you had been told the speed in mph, you would have had to convert this to $\mathrm{m\,s^{-1}}$ before using it.

Example 2

Calculate the constant acceleration in Example 1.

Solution

Now you know $u = 0$, $v = 13$ and $t = 6$ and you need a.

Choose $\quad v = u + at$

$$13 = 0 + a \times 6$$

$$a = 13 \div 6$$

The acceleration is $2.17\,\mathrm{m\,s^{-2}}$ correct to 3 s.f.

> **Common mistake:** Always make sure your units are compatible.

> Look which quantity is not given nor required; in this question it is s. Choose the equation that does not contain that quantity.

> Substituting the values you know.

Example 3

Starting at $25\,\mathrm{m\,s^{-1}}$, a car slows down (decelerates) at $2\,\mathrm{m\,s^{-2}}$.

i   How long does it take to stop?

ii How long does it take to travel 100 m from its starting point?

Solution

i   *A deceleration (or retardation) of $2\,\mathrm{m\,s^{-2}}$ is an acceleration of $-2\,\mathrm{m\,s^{-2}}$*

You know $u=25$, $v=0$, $a=-2$ and you need t.

Choose $v = u + at$

$\qquad 0 = 25 - 2t$

The time to stop is 12.5 s

> In this question s is not given nor required so choose the equation that does not involve s.

ii *Now you know $u=25$, $a=-2$ and $s=100$, and you need t.*

Choose $\qquad s = ut + \frac{1}{2}at^2$

Giving $\qquad 100 = 25t - 1t^2$

$\qquad\qquad t^2 - 25t + 100 = 0$

Factorising $\qquad (t-5)(t-20) = 0$

$\qquad\qquad t = 5 \text{ or } t = 20$

> Alternatively you can use your calculator to solve this equation.

The car had stopped after 12.5 s so $t=20$ is rejected.

The answer is that the car takes 5 s to travel 100 m.

Example 4

A car travels along a straight road ABC. AB is 27 m and BC is 100 m. The car starts from rest at A, accelerates uniformly at $1.5\,\mathrm{m\,s^{-2}}$ to B and then travels at constant velocity to C. How long does this total journey take?

Solution

Start by putting the information on a diagram:

A $\xrightarrow{\quad}$ $a_1 = 1.5$ B $\xrightarrow{\quad}$ $a_2 = 0$ C

$s_1 = 27$ $s_2 = 100$

$t_1 = ?$ $t_2 = ?$

$\xrightarrow{} u_1 = 0$ $\xrightarrow{} v_1 = u_2 = ?$ $\xrightarrow{} v_1 = u_2$

> There are two parts to the motion so use
> u_1, v_1, s_1, t_1, a_1 for A to B and
> u_2, v_2, s_2, t_2, a_2 for B to C.
> Notice that $v_1 = u_2$ the velocity at B and at C.

First find the time, t_1 s to accelerate (between A and B)

Use $\qquad s = ut + \frac{1}{2}at^2$

You know $s_1 = 27$, $u_1 = 0$ and $a = 1.5$. You need t_1.

$27 = 0 + \frac{1}{2} \times 1.5 \times t_1^2$

$t_1^2 = 2 \times 27 \div 1.5$

$t_1 = \sqrt{36} = 6$

> **Common mistake:** Remember the square root

For the second stage, from B to C, you need the constant speed.

This is the value of v at B, the end of the first stage.

> You know $u_1 = 0$, $a_1 = 1.5$, $s_1 = 27$ and you need v_1.

$v^2 = u^2 + 2as$

Using $\qquad v_1^2 = 0 + 2 \times 1.5 \times 27$

$v_1 = \sqrt{81} = 9$

The time taken at constant speed equals $\dfrac{distance}{speed}$ or $\dfrac{s_2}{v_1}$.

$$t_2 = 100 \div 9$$
$$= 11\tfrac{1}{9}$$

The total time taken is $t_1 + t_2 = 6 + 11\tfrac{1}{9}$ s or 17.1 s correct to 3 s.f.

Example 5

A motorcyclist travels along a straight track. Her motion is modelled using the assumption that her acceleration is constant. She travels from A to B in 5 s and from B to C in the next 5 s. AB is 55 m and BC is 115 m.

i Calculate her speed at A and the acceleration.
ii Calculate the distance travelled according to this model in the next 5 s.
iii Why might this model not be appropriate for the 5 s after that?

Solution

i There are two unknowns, so set up a pair of simultaneous equations.

You need to use u for the initial speed at A, so you need to look at the journeys from A to B and from A to C.

From A to B $s = ut + \tfrac{1}{2}at^2$ becomes $55 = 5u + \tfrac{1}{2}a \times 5^2$ or $5u + 12.5a = 55$

From A to C $s = ut + \tfrac{1}{2}at^2$ becomes

$55 + 115 = 10u + \tfrac{1}{2}a \times 10^2$ or $10u + 50a = 170$

Solve the simultaneous equations) to give $u = 5$ and $a = 2.4$ ◄────

So her speed at A is 5 ms⁻¹ and her acceleration is 2.4 ms⁻².

$$10u + 50a = 170 \quad \blacktriangleleft$$
$$10u + 25u = 110$$
$$25a = 60$$
$$a = 2.4$$
$$u = 5$$

ii When $t = 15$ s
$$s = ut + \tfrac{1}{2}at^2 = 5 \times 15 + \tfrac{1}{2} \times 2.4 \times 15^2 = 345$$
Distance travelled beyond C is $345 - 170 = 175$ m.

iii At the end of 15 s, the speed of the motorcyclist is
$$v = u + at = 5 + 2.4 \times 15 = 41 \text{ ms}^{-1}.$$
This is 147.6 km per hour (about 92 mph).

By now she would be travelling very fast and so it is more likely that she has not maintained the same constant acceleration.

Common mistakes:
- For the second part of Example 4, you could have used $v = u + at$ with your answer for t_1 to find v, but a mistake in t_1 would have led to another mistake in v. It is better to use the information given in the question if you can.
- Be especially careful in questions where the direction of motion changes. The displacement s would then not be the same as the distance travelled.

Common mistake: You cannot use the same letter for the initial speed for AB and the initial speed for BC.

Use the solve facility on your calculator if you have it.

Multiply the first equation by 2 and subtract from the second equation.

Common mistake: Include a value from your previous work or a new calculation to support your answer. Do not just give a vague answer like 'she would be going too fast'.

Test yourself

Before you start these questions, cover Question 1 and write down the following *suvat* equations from memory. Do all of them before you check your answers.

Make sure you learn any you get wrong and try this test again after doing Question 1.

- a An equation involving *s, t, u, v*
- b An equation involving *a, t, u, v*
- c An equation involving *a, s, u, v*
- d An equation involving *a, s, t, v*
- e An equation involving *a, s, t, u*

1 For each of the following decide which *single* equation is the most appropriate:

A $v = u + at$

B $s = \frac{1}{2}(u+v)t$

C $s = ut + \frac{1}{2}at^2$

D $v^2 = u^2 + 2as$.

i A van travelling at $12\,\text{m}\,\text{s}^{-1}$ stops in $4\,\text{s}$. How far does it travel in that time?

ii A bus stops in a distance of $100\,\text{m}$ from a speed of $10\,\text{m}\,\text{s}^{-1}$. What is its acceleration?

iii A ball is dropped from a window at a height of $15\,\text{m}$ and accelerates at $9.8\,\text{m}\,\text{s}^{-2}$. How long does it take to reach the ground?

iv What is the speed of the ball in **iii** when it has fallen $10\,\text{m}$?

2 A light aircraft lands with a speed of $30\,\text{m}\,\text{s}^{-1}$ and takes $20\,\text{s}$ to come to rest. Assuming constant acceleration, three of the following statements are true and one is false. Which one is false?

A The aeroplane travels $300\,\text{m}$ before coming to rest.

B The acceleration of the aeroplane is $-1.5\,\text{m}\,\text{s}^{-2}$

C The aeroplane travels half the distance in the first $10\,\text{s}$.

D The speed of the aeroplane is halved after $10\,\text{s}$.

3 A train accelerates from rest at $0.2\,\text{m}\,\text{s}^{-2}$ for $4000\,\text{m}$. How long does it take and how fast is it travelling?

A $200\,\text{s}, 40\,\text{m}\,\text{s}^{-1}$

B $200\,\text{s}, 20\,\text{m}\,\text{s}^{-1}$

C $20\,\text{s}, 40\,\text{m}\,\text{s}^{-1}$

D $20\,\text{s}, 4\,\text{m}\,\text{s}^{-1}$

4 A car starts at rest and accelerates at $2\,\text{m}\,\text{s}^{-2}$ for $4\,\text{s}$. It then travels at constant speed for $10\,$minutes. How far has it travelled in this time?

A $76\,\text{m}$

B $196\,\text{m}$

C $4800\,\text{m}$

D $4816\,\text{m}$

5 Kirsty runs at a constant speed of $8\,\text{m}\,\text{s}^{-1}$. She gives a $15\,\text{m}$ head start to Jonathan who starts from rest and has a constant acceleration of $1\,\text{m}\,\text{s}^{-2}$. Which equation gives the time at which Kirsty catches Jonathan?

A $8t = t^2$

B $8t = \frac{1}{2}t^2$

C $8t = \frac{1}{2}t^2 + 15$

D $8t + 15 = \frac{1}{2}t^2$

Full worked solutions online

Exam-style question

A car is travelling along a straight road with an initial speed of $11\,\text{m}\,\text{s}^{-1}$ when it starts to slow down at a traffic light. Its deceleration is constant. The car stops after $3\,\text{s}$, waits for $5\,\text{s}$ and then accelerates uniformly to its original speed of $11\,\text{m}\,\text{s}^{-1}$ in $4\,\text{s}$.

i Calculate the distance travelled by the car during these $12\,\text{s}$.

ii How much less time would the car have taken to travel this distance if it had maintained the speed of $11\,\text{m}\,\text{s}^{-1}$ throughout the $12\,\text{s}$?

Short answers on page 161

Full worked solutions online

Vertical motion under gravity

Key facts

- The acceleration due to gravity, $g\,\mathrm{m\,s^{-2}}$, varies around the world. In this book it is assumed that all questions arise in a place where its value is $9.80\,\mathrm{m\,s^{-2}}$ vertically downwards; this is usually written as $9.8\,\mathrm{m\,s^{-2}}$.
- $v = 0$ at the highest point of the motion.
- Always draw a diagram and decide in advance where your origin is and which way is positive. $s = 0$ is the origin. Whatever the position, v is positive in the positive direction. a is $+9.8$ when downwards is positive and -9.8 when upwards is positive.

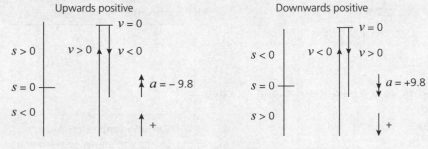

- In problems where the motion does not begin at the origin so that $s = s_o$ when $t = 0$, replace s in each equation with $(s - s_o)$

Worked examples

Example 1

A ball is projected upwards at $4.9\,\mathrm{m\,s^{-1}}$.

  I Find the time it takes to reach its maximum height.

  ii Find the maximum height.

Solution

  i *The ball reaches its maximum height when its velocity is zero.*

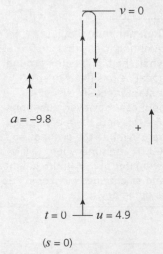

First you need to find the time this occurs.
Take upwards as positive and $t = 0$ when it is projected.
You know $u = 4.9$, $a = -9.8$ and $v = 0$ and you need t.

In this part of the question s is not given nor required so choose the equation that does not involve s.

Choose $\quad v = u + at$

$$0 = 4.9 + (-9.8)t$$
$$9.8t = 4.9$$
$$t = 0.5$$

ii Now take $s = 0$ when $t = 0$.

To find the maximum height, you need to find s when $t=0.5$
$u = 4.9$, $a = -9.8$, $s = ?$
Choose $s = ut + \frac{1}{2}at^2$

$$s = 4.9 \times 0.5 + \frac{1}{2} \times \left(-9.8\right) \times \left(0.5\right)^2$$
$$= 2.45 - 1.225 = 1.225$$

The maximum height is 1.23 m (3 s.f.)

> Displacement is measured from the point at which the ball is projected

> You could also use $v = 0$ instead of $t = 0.5$

Example 2

A ball is projected vertically upwards from a height of 1 m and reaches its maximum height after 2 s. Find:

i   the initial velocity of the ball

ii its velocity when it hits the ground.

Solution

i Take upwards to be positive.

You know $v = 0$ when $t = 2$ and $a = -9.8$ and you are asked to find u,

Choose $v = u + at$

$$0 = u - 9.8 \times 2$$
$$u = 19.6$$

The initial velocity is 19.6 m s^{-1} upwards.

ii To find the velocity when the ball hits the ground, you need v when $s = -1$ assuming $s = 0$ when $t = 0$.

You also need to use the value of u that you have just found.

Choose $v^2 = u^2 + 2as$

$$v^2 = 19.6^2 + 2 \times (-9.8) \times (-1)$$
$$v^2 = 403.76$$
$$v = -\sqrt{403.76}$$
$$= -20.1$$

When the ball hits the ground it has a velocity of 20.1 m s^{-1} downwards.

> **Common mistake:** You are asked for the velocity so you have to give both the speed (19.6 m s^{-1}) and the direction (upwards).

> v is negative means that the ball is moving downwards.

> **Common mistake:** It is tempting to say that $v = 0$ when the ball hits the ground because it is likely to stop instantaneously even if it bounces. However, questions like this actually mean 'just before it hits the ground', so $v \neq 0$. The time when $v = 0$ is when it is at maximum height and so changes from going up to going down.

Example 3

A ball is dropped from a height of 8 m, hits the ground and rebounds at half the speed. How high does it bounce?

Solution

First find the speed of the ball when it hits the ground. For this part of the motion,

- take downwards to be positive
- take the point where it is dropped as origin.

> It is tempting to assume that the first bounce is half the height, but you might be wrong. It is better to work out the answer properly.

You know $u = 0$, $a = 9.8$, $s = 8$, and want to find v

Choose:
$$v^2 = u^2 + 2as$$
$$= 0 + 2 \times 9.8 \times 8$$
$$v = \sqrt{156.8}$$
$$= 12.5219...$$

Rebound speed $= \frac{1}{2} \times 12.5219$
$$= 6.2609... \, \text{ms}^{-1}$$

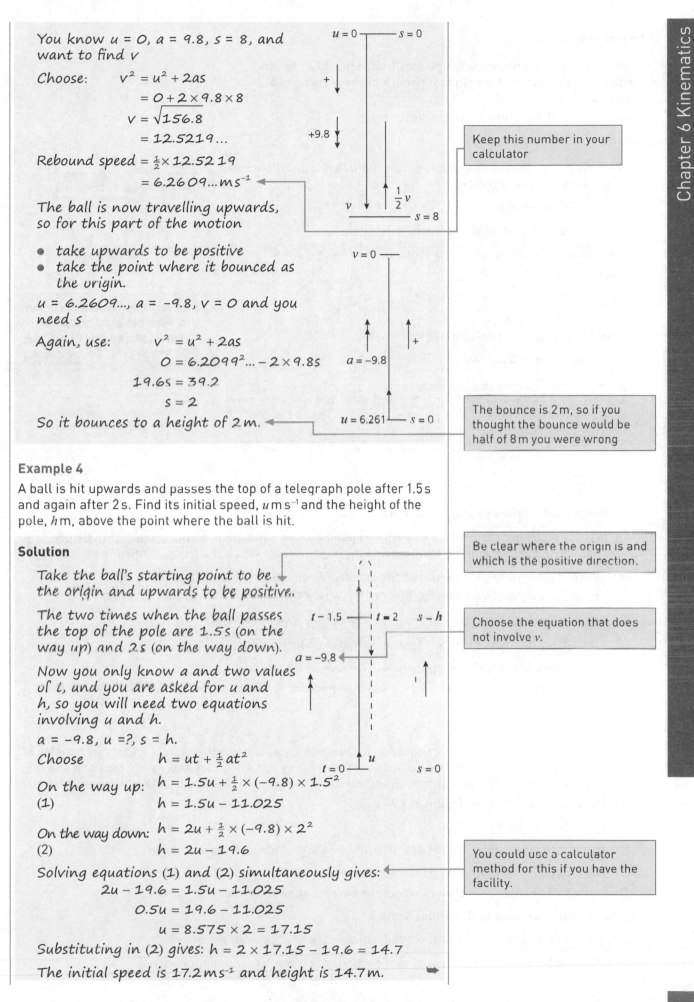

Keep this number in your calculator

The ball is now travelling upwards, so for this part of the motion

- take upwards to be positive
- take the point where it bounced as the origin.

$u = 6.2609...$, $a = -9.8$, $v = 0$ and you need s

Again, use:
$$v^2 = u^2 + 2as$$
$$0 = 6.2099^2... - 2 \times 9.8s$$
$$19.6s = 39.2$$
$$s = 2$$

So it bounces to a height of 2 m.

$a = -9.8$

$u = 6.261$ $s = 0$

The bounce is 2m, so if you thought the bounce would be half of 8m you were wrong

Example 4

A ball is hit upwards and passes the top of a telegraph pole after 1.5 s and again after 2 s. Find its initial speed, $u \, \text{m s}^{-1}$ and the height of the pole, h m, above the point where the ball is hit.

Solution

Take the ball's starting point to be the origin and upwards to be positive.

Be clear where the origin is and which is the positive direction.

The two times when the ball passes the top of the pole are 1.5 s (on the way up) and 2 s (on the way down).

$t = 1.5$ $t = 2$ $s = h$

$a = -9.8$

Choose the equation that does not involve v.

Now you only know a and two values of t, and you are asked for u and h, so you will need two equations involving u and h.

$a = -9.8$, $u = ?$, $s = h$.

Choose
$$h = ut + \tfrac{1}{2}at^2$$

$t = 0$ u $s = 0$

On the way up: $h = 1.5u + \tfrac{1}{2} \times (-9.8) \times 1.5^2$
(1) $h = 1.5u - 11.025$

On the way down: $h = 2u + \tfrac{1}{2} \times (-9.8) \times 2^2$
(2) $h = 2u - 19.6$

Solving equations (1) and (2) simultaneously gives:
$$2u - 19.6 = 1.5u - 11.025$$
$$0.5u = 19.6 - 11.025$$
$$u = 8.575 \times 2 = 17.15$$

You could use a calculator method for this if you have the facility.

Substituting in (2) gives: $h = 2 \times 17.15 - 19.6 = 14.7$

The initial speed is 17.2 ms⁻¹ and height is 14.7 m.

Example 5

A small ball is thrown vertically upwards from a height of 1 m and takes 2 seconds to reach its highest point, h m above its point of projection.

How far above the ground is the greatest height?

Solution

The diagram shows the path of the ball. Assume upwards is positive so the acceleration is

$$-g = -9.8 \text{ ms}^{-1}$$

At the top: $v = 0$ and $t = 2$ and $s = h$.

You do not know u and you need h.

Choose $s = vt - \frac{1}{2}at^2$

Then
$$h = 0 - \frac{1}{2} \times (-9.8) \times 2^2$$
$$h = 19.6$$

The height above the ground is
$$h + 1\text{m} = 20.6\text{m}$$

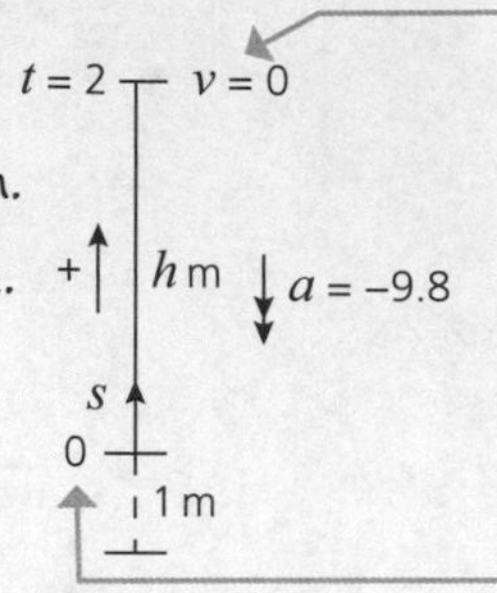

At the highest point $v = 0$

$t = 2$ $v = 0$

h m $a = -9.8$

s

0

1 m

Be very clear where the origin is. Take into account that the point of projection is 1 m above the ground.

You can take s as the distance from the ground, but then $s = 1$ when $t = 0$ so the equation to use is $s = 1 + vt - \frac{1}{2}at^2$ or $s - 1 = vt - \frac{1}{2}at^2$. In this case $t = 2$ gives $s = 20.6$ directly.

Test yourself

TESTED

This information applies to Questions 1 to 3.

A ball is thrown upwards from a window with a speed of 3 ms^{-1} and lands on the ground 15 m below. Take downwards to be positive and the level outside the window from which the ball is thrown to be the origin.

Three of these statements are false and one is true. Which one is the true statement?

1 A   The distance travelled by the ball before it lands is 15 m.

   B   The displacement of the ball is positive when it is above the window.

   C   The ball's velocity is $+10 \text{ ms}^{-1}$ at some point during the motion.

   D   The velocity of the ball is negative when it is below the window.

2 Which one of these is an equation which will give the time, t, taken for the ball to reach the ground?

   A $4.9t^2 - 3t - 15 = 0$ B $4.9t^2 + 3t - 15 = 0$

   C $4.9t^2 + 3t + 15 = 0$ D $9.8t^2 - 3t - 15 = 0$.

3 When you have solved the correct equation in Question 2 you can use the results and the symmetry of the motion to give you even more information. Three of the following statements are true and one is false. Solve the correct equation from Question 2 and use your answers to decide which one is false.

   A   The ball is at maximum height at 1.04 s

   B   The ball is in the air for 2.08 s

   C   The ball reaches the ground 1.47 s after passing the window on the way down

   D   The ball passes the window on the way down 0.61 s after it was thrown up.

Try to think of another way of working out the correct answer for each statement.

This information applies to Questions 4 and 5

Starting from rest, a rocket is fired vertically upwards with acceleration 25 ms^{-2}.

After 0.8 seconds there is no more fuel so it continues to move freely under gravity.

4 Calculate the maximum height of the rocket.

  A 8 m   B 16 m   C 20.4 m   D 28.4 m

5 From the moment when the rocket has no more fuel, it takes T s to return to earth. Which of these is a correct equation for T?

  A $4.9T^2 - 20T - 8 = 0$ B $4.9T^2 + 20T - 8 = 0$

  C $4.9T^2 - 20T + 8 = 0$ D $4.9T^2 + 20T + 8 = 0$

Full worked solutions online

CHECKED ANSWERS

Exam-style question

A particle, P, is projected vertically upwards at $21\,\text{m s}^{-1}$ from a point O on the ground.
i Calculate the maximum height of P.
When P is at its highest point, a second particle, Q, is projected upwards from O at $15\,\text{m s}^{-1}$.
ii Show that P and Q collide 1.5 s later and determine the height above the ground that this takes place.

Short answers on page 161

Full worked solutions online

CHECKED ANSWERS

Chapter 7 Forces and Newton's laws of motion

About this topic

Forces are absolutely essential to Mechanics. There are various types of mechanical force and it is essential to know which of them are acting in any situation and to be able to represent them on a diagram, showing their magnitudes and directions, for example when two objects are in contact or connected together.

The basic rules governing the effects of forces are expressed as Newton's laws of motion. Newton's 3rd law, involving action and reaction, is used when two objects are in contact. Newton's 1st and 2nd laws deal with the relationship between force and motion.

Before you start, remember ...

- Forces are vectors. They can be represented by a magnitude (size) and a direction given by an arrow.
- Displacement, velocity, acceleration and force are vectors so each one has a magnitude (size) and a direction.
- Mass, length, speed and time are scalars with magnitude only.
- Vectors can be added and subtracted and multiplied by a scalar.
- Forces and acceleration are vectors. They can be represented by a magnitude (size) and a direction given by an arrow.
- The *suvat* equations for motion with constant acceleration.
- Drawing force diagrams.
- Solving simultaneous equations.
- Newton's 2nd law: Resultant force = mass × acceleration, or $F = ma$.

Key facts

- **Newton's laws of motion**
  1. Every particle continues in a state of rest or of uniform motion in a straight line unless acted on by a resultant external force.
  2. The change in motion is proportional to the force.
  3. When an object exerts a force on another there is always a reaction of the same kind which is equal, and opposite in direction, to the acting force.
- **Types of force**
  i   Forces on an object due to contact with the surface of another.
      - **Friction** *always* opposes any tendency to slide.
      - The **normal reaction** is always perpendicular to the surfaces and to any friction.
      These are the two components of the reaction between the surfaces in contact.

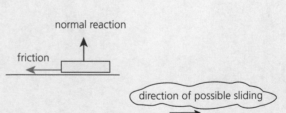

---

ii Forces in a joining rod or string:

Tension

The tension and compression forces shown act on the objects attached to the ends

Thrust or compression
(rod only)

iii The tension in a string has the same magnitude on each side of a smooth light pulley but a different direction.

T

T

iv The horizontal forces on a wheeled vehicle are usually reduced to three possible forces: resistance, driving force and braking force.

v The **weight** of an object is the force of gravity pulling it towards the centre of the earth. Weight$=mg$ vertically downwards. The weight of an object is represented by one force acting through its centre of mass.

Resistance

Driving force

Braking force

● **Commonly used modelling terms**

inextensible	does not vary in length
light	negligible mass
negligible	small enough to ignore
particle	negligible dimensions
smooth	negligible friction
uniform	the same throughout

Forces

Worked examples

Example 1

Draw diagrams showing the forces acting on:

i a computer mouse which is being moved along a table by a horizontal force P.

ii the table due to the presence of the mouse.

Solution

i *forces on the mouse*

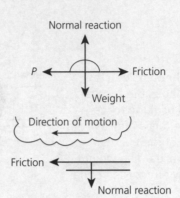

ii *forces of mouse on table*

The friction and normal reaction forces acting on the table are equal and opposite to those of the table on the mouse.

> **Common mistake:** Notice that the weight of the mouse and the force P pushing it are present only in the diagram for the mouse.
>
> Because the mouse is moving horizontally, the normal reaction between it and the table is equal to its weight. However, the vertical reaction between any two objects is *not necessarily* equal to the weight of the one on top.

Example 2

The diagram shows a pile of three blocks, A, B and C at rest on a horizontal table. Their weights are W_1, W_2 and W_3 respectively. Draw diagrams to show the forces acting on each of the blocks.

Solution

The diagram shows the forces on the three blocks.

Forces on block A

Forces on block B

Forces on block C

  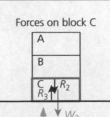

> Notice that the force that block B exerts on block A, R_1, has the same size as the force that block A exerts on block B but is in the opposite direction.

> Similarly, the force that block C exerts on block B, R_2, has the same size as the force that block B exerts on block C but is in the opposite direction.

Example 3

The diagram below shows a block on a rough horizontal table. A string, which is in tension, is attached to one corner. The block is stationary. Draw a diagram to show the forces acting on the block.

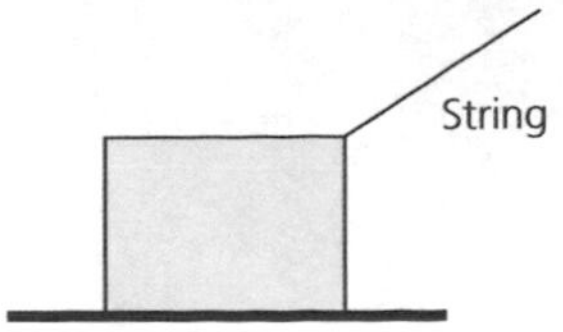

String

Solution

The diagram shows the solution

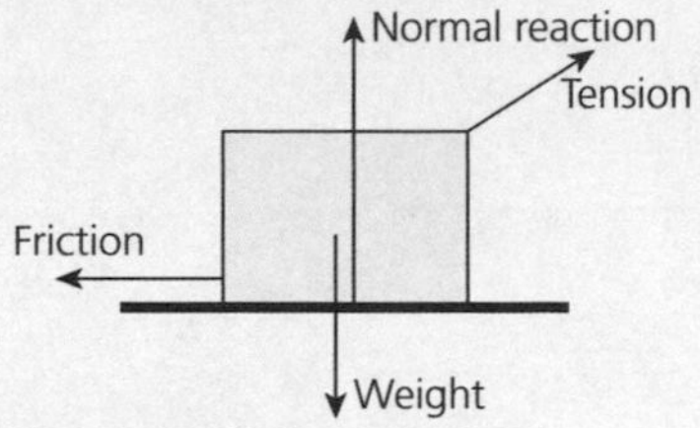

Common mistake: A common mistake in a question like this is to miss out the frictional force. You are told that the table is rough and that means that there may be a frictional force but it is not certain.

The tension in the string produces a force to the right (as well as upwards).

You are also told that the block is stationary. So there must be a force to the left to oppose it and that has to be friction.

Example 4

A particle of mass m is on a smooth uniform slope that makes an angle of α to the horizontal. It is held at rest by a light string which makes an angle of β with the slope. The tension in the string is T.
Draw a diagram showing all the forces acting on the particle.

Notice that you are told that the slope is smooth. That tells you that there can be no frictional force.

Solution

There are three forces acting on the particle as shown in the figure.

- *its weight mg*
- *the normal reaction of the slope, R*
- *the tension in the string, T.*

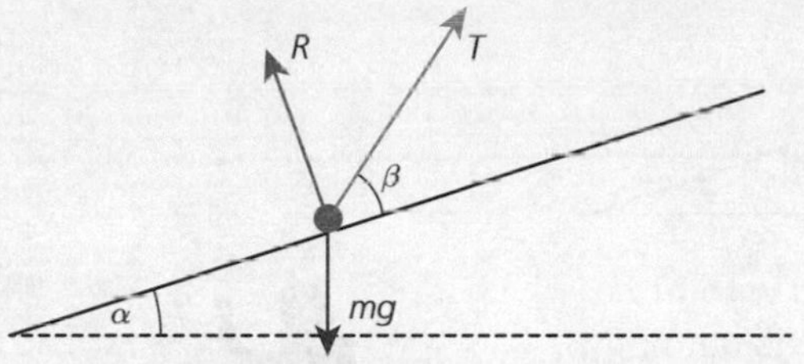

Hint: Notice that no units are given in this question. In such cases it is assumed that the various quantities are measured in a consistent set of units, such as the S.I. system.

Example 5

A girl attached to a rope is being lowered down a rock face. The rope passes through a smooth pulley at the top of the cliff and is held by a rock climber. The girl has her feet on a ledge and is momentarily at rest.

Draw a diagram to show the forces acting on:

i the girl

ii the rock climber.

M_1 is her mass

T is the tension in the rope.

N_1 is the normal reaction of the ledge.

F_1 is the friction between her and the ledge

M_2 is his mass.

N_2 is the normal reaction of the ground.

F_2 is the friction between him and the ground.

Solution

i *Forces on the girl*

ii *Forces on the rock climber*

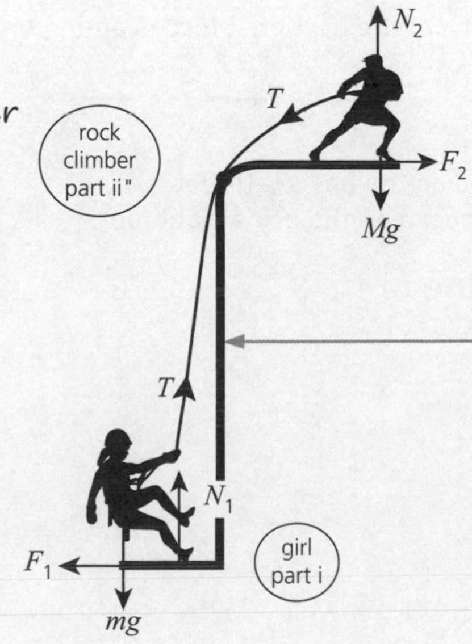

All forces are in newtons.

The tension in the rope is the same in both cases because the pulley is smooth

Test yourself

Questions 1 and 2 are about a box at rest on a rough plane.

1 Which of these is the correct diagram showing the forces acting the box?

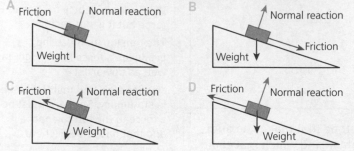

2 Which of these is the correct diagram showing the forces the box exerts on the plane?

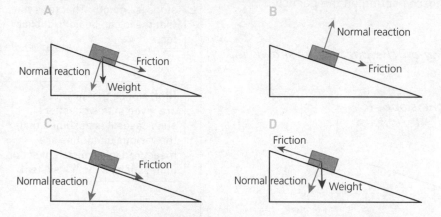

Questions 3 and 4 are about a boy who pushes a skateboard with his right foot while his left foot pushes against level ground. He is gaining speed.

forwards

3 One of the following statements MUST be true, the others may be false.
   Which one must be true?

   A The vertical reaction between the boy's right foot and the skateboard is equal to his weight.

   B There is a forwards frictional force on the boy's left foot.

   C The forces on the boy are in equilibrium.

   D The boy's weight is equally distributed between the ground and the skateboard.

4 Which diagram shows the directions of the frictional forces acting on the boy:

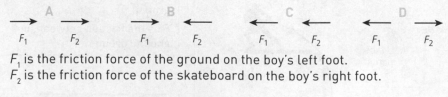

F_1 is the friction force of the ground on the boy's left foot.
F_2 is the friction force of the skateboard on the boy's right foot.

Full worked solutions online

CHECKED ANSWERS

Exam-style question

A student is moving a box of books of mass m kg to her lodgings.
First she pushes it along rough level ground with a force P as shown in the diagram

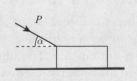

i   Draw a diagram to show the forces acting on the box. Is the normal reaction with the ground equal to the weight of the box?

When the student reaches a ramp, she decides to pull the box with a rope making an angle of β to the horizontal as in the diagram to the right. The ramp also is rough and makes an angle of θ with the horizontal.

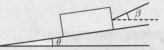

ii   Draw a diagram to show the forces now acting on the box.

Short answers on page 161

Full worked solutions online

CHECKED ANSWERS

Newton's 1st law of motion

REVISED

Key facts

● **Newton's 1st law of motion**
Every object continues in a state of rest or uniform motion in a straight line unless it is acted on by a resultant force.

● **Equilibrium**
A particle is in equilibrium when it is stationary or travelling in a straight line at constant speed.
  o   For forces acting in the same line, the resultant force on the particle is zero.
  o   For forces given in two perpendicular directions, the resultant force in each direction is zero.
  o   For forces given in vector notation, the total force is the zero vector.

● **Weight**
The weight of an object is the force of gravity pulling it towards the centre of the earth. Weight $= mg$ vertically downwards. The weight of an object is represented by one force acting through its centre of mass.

● **S.I. units**

length	metre	m
time	second	s
velocity	metres per second	$m\,s^{-1}$
acceleration	metres per second per second	$m\,s^{-2}$
mass	kilogram	kg
force	newton	N

1 newton is the force required to give a mass of one kilogram an acceleration of $1\,m\,s^{-2}$.
1000 newtons $= 1$ kilonewton (kN)

Newton's 1st law

When the forces on an object are balanced and have *no resultant* they are said to be *in equilibrium*. Newton's 1st law says that in that case the velocity does not change. The object could be stationary (at rest) or it could have constant velocity. Constant velocity means that both the magnitude and the direction of the velocity remain unchanged. The object must have constant speed and must be moving in a straight line.

Worked example

Example 1

In each of the following cases state whether the forces on the object are in equilibrium.

   i   A car is cruising along a straight motorway at 60 mph.

   ii A cable car on a straight wire pulls in to its terminal.

   iii The minute hand on my watch rotates at a steady speed.

   iv A kestrel's eye is fixed as it hovers over the ground.

Solution

   i   *A cruising car is assumed to have a constant speed and the road is straight so the forces on the car as a whole must be in equilibrium.*

   ii *The cable car must be slowing down as it pulls into the terminal so its speed is not constant. The forces are not in equilibrium.*

   iii *The minute hand is not travelling in a straight line so the forces are not in equilibrium.*

   iv *As a bird of prey, the kestrel needs a steady eye. When it is stationary the forces acting on it are in equilibrium.*

> **Common mistake:** Take care **not** to assume that a body is necessarily at rest when the forces acting on it are said to be in equilibrium.

Whenever either its speed or its direction of motion changes, an object has an acceleration. It could be travelling in a straight line with variable speed, or moving with constant speed along a curve, or both speed and direction could be changing.

In order to produce any acceleration, there must be a resultant force acting on the object in the same direction as the acceleration.

Worked examples

Example 2

A 60 kg parachutist has reached terminal velocity so is falling at a constant speed. What is the air resistance acting on the parachutist?

Solution

The parachutist is falling at a constant speed so, assuming this is in a straight line, the forces acting are in equilibrium. The air resistance, R, is equal to the weight of the parachutist, mg.

> A diagram of forces is useful even in simple cases. You can simplify the drawing to a small circle with two arrows.

Air resistance $= 60 \times g\,\text{N}$
$\qquad\qquad\quad = 588\,\text{N}$

Example 3

In each case, state whether the force of the person on the wall is equal to, less than or greater than the weight of the person.

i   You are sitting on a low wall with your feet on the ground.

ii A small child is sitting on the wall with feet dangling.

iii You jump down onto the wall and it is the moment when you are landing.

Solution

i   *You are at rest so the forces acting on you are in equilibrium. These are your weight downwards and the upwards forces due to contact with the wall and also the ground. Hence the force of the wall on you must be less than your weight. The equal force you apply to the wall is therefore less than your weight.*

ii *This time there is no force on the ground so the force on the wall is equal to the weight of the child.*

iii *When you land the upward force acting on you must be greater than your weight in order to slow you down. Hence the equal and opposite downward force on the wall must also be greater than your weight.*

> Wherever there is contact with the ground, there will be a normal reaction.

> This is a situation with acceleration so the forces are not in equilibrium.

Example 4

A tractor of mass 3000 kg is towing a trailer of mass 1000 kg at a steady speed along a straight horizontal road.

There is a driving force of 800 N.
There are also resistance forces of 500 N on the tractor and 300 N on the trailer.

i   Draw clearly labelled diagrams to show the forces on the tractor and on the trailer.

ii Which of these forces are internal forces when the whole system is considered?

iii Why is the driving force (800 N) equal to the sum of the resistances to motion (500 N + 300 N)?

iv Find the vertical reaction forces and the tension in the towbar.

➡

Solution

i **Forces on trailer**   **Forces on tractor**

R_1 N and R_2 N are the normal reactions of the road on the tractor and trailer respectively. TN is the tension in the towbar.

> It is important to attach the forces to the tractor or the trailer. It matters where the resistances act.

> Notice that the *weights* of the tractor and trailer are each given as mg N.

ii The tension, TN, in the towbar is an internal force when the whole system is considered. The force of tractor on trailer is equal and opposite to the force of trailer on tractor. (Newton's 3rd law)

iii By Newton's 1st law, the steady speed means that the forces on the whole system are in equilibrium, so the horizontal driving force 800 N is correctly given to be equal to the sum of the horizontal resistances.

> Consider the tractor and trailer together as a single object. Make sure you state that the system is in equilibrium.

iv In addition, the forces on each of the tractor and trailer independently are in equilibrium.

For the tractor: $R_1 = 3000g$ > Vertical forces balance

and $800 = T + 500$ > Horizontal forces balance

so $T = 300$

For the trailer: $R_2 = 1000g$ > Vertical forces balance

$T = 300$ as before.   > Horizontal forces balance

Example 5

In this question **i** is a horizontal unit vector and **j** is an upward vertical unit vector.

A particle of mass 3 kg is in equilibrium under the action of its weight and two other forces $\mathbf{F}_1 = a\mathbf{i} + 7\mathbf{j}$ and $\mathbf{F}_1 = 2\mathbf{i} + b\mathbf{j}$.

i Write the weight **W** in the from $\mathbf{W} = x\mathbf{i} + y\mathbf{j}$

ii Find the values of the constants a and b.

Solution

i The weight of the particle acts vertically downwards and has no sideways component so it is in the negative **j** direction.

> Remember that weight = mg.

$\mathbf{W} = 0\mathbf{i} - 3g\,\mathbf{j} = -29.4\,\mathbf{j}$

ii The particle is in equilibrium so the total force = 0

$\mathbf{F}_1 + \mathbf{F}_1 + \mathbf{w} = 0$

$(a\mathbf{i} + 7\mathbf{j}) + (2\mathbf{i} + 6\mathbf{j}) - 29.4\mathbf{j} = 0$

In the **i** direction $a + 2 = 0$ so $a = -2$

In the **j** direction $7 + b - 29.4 = 0$ so $b = 22.4$

> Build a vector equation using the fact that the total force is zero. It can be solved by considering the two directions separately.

Test yourself

1 In each of these situations decide whether, during the situation described, the forces acting on the object are in equilibrium:

A always

B never

C some of the time but not all of it

D there is insufficient information to tell.

i A book is lying on a table.

ii A sky diver is in free fall before reaching terminal velocity.

iii A lift in Canary Wharf rises without stopping from ground level to the fiftieth floor.

iv A child is sitting in a car on a fairground roundabout which is rotating with constant speed.

v A person is sitting on a bus.

2 Three of the following statements are true and one is false. Which one is false?

A You are in a lift. When it starts moving the force between you and the floor is not equal to your weight.

B A 1.5 kg bag of rice has a weight of 14.7 N.

C When something is moving, there must be a force making it move.

D When you are sitting on a chair, the force on the seat of the chair is normally less than your weight.

3 A small box is in equilibrium on a horizontal table. It is pulled by a string with tension T. Three of the following statements are true and one is false.
Which one is false?

A The friction force, F, is in the wrong direction in the diagram.

B R is the normal reaction of the table on the box.

C $R = mg$

D The magnitude of the tension, T, is greater than the magnitude of F

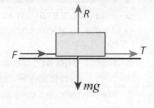

4 A tractor is pulling a trailer with no brakes along level ground using a simple towbar. The tractor brakes suddenly. Draw for yourself a rough sketch showing the forces acting on the tractor and trailer at this moment. Take care to indicate the directions of all the forces.
Three of the following statements are true and one is false. Which one is false?

A The forces are not in equilibrium.

B The tractor and trailer slow down at the same rate.

C The tension in the tow bar must be zero.

D The braking force acts only on the tractor.

5 A particle is in equilibrium under the action of three forces. $\mathbf{F_1} = 12\mathbf{i} - 5\mathbf{j}$, $\mathbf{F_2} = 3(\mathbf{i} + \mathbf{j})$ and $\mathbf{F_3}$. Which of the following is the correct expression for $\mathbf{F_3}$?

A $\mathbf{F_3} = 15\mathbf{i} - 2\mathbf{j}$

B $\mathbf{F_3} = -15\mathbf{i} + 4\mathbf{j}$

C $\mathbf{F_3} = -13$

D $\mathbf{F_3} = -15\mathbf{i} + 2\mathbf{j}$

Full worked solutions online

Exam-style question

The diagram shows two boxes of mass 5 kg and 7 kg standing together on a rough horizontal surface. Ruth tries to push the boxes by applying a horizontal force of 30 N as shown in the diagram. There are frictional forces of $5k$ N and $7k$ N acting on the two boxes. The boxes do not move.

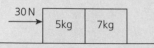

i Draw separate diagrams for each box showing all the forces acting on them.
ii Find the magnitude of the frictional forces acting on the 5 kg box.
iii Find the normal reaction between the 7 kg box and the surface.
iv Find the contact force between the two boxes.

Short answers on page 161

Full worked solutions online

CHECKED ANSWERS

Applying Newton's 2nd law along a line

REVISED

Key facts

● **Newton's 2nd law:**
  o The resultant force is equal to the mass of the object multiplied by the acceleration.
  o For motion in a straight line this is often written as $F = ma$
  o The acceleration is in the same direction as the resultant force.
  o The equation obtained is often referred to as the 'equation of motion'
  o Force and acceleration are vector quantities, so the equation can be written **F** = m**a**
  o Newton's 1st law is a special case of his 2nd with both **F** and **a** equal to zero.
● **The S.I. units** are:

force:	newtons, N
acceleration:	$m\,s^{-2}$

Hint: Acceleration and force are different quantities so it is helpful to use different types of arrow for these: force: → acceleration: →→.

Newton's 2nd law

A force of 1 N gives a mass of 1 kg an acceleration of $1\,m\,s^{-2}$. This results in the important equation:

$$F = ma$$

Any object falling to earth has an acceleration of $g\,m\,s^{-2}$ vertically downwards, so the force of gravity acting on it is mg N. This is its **weight**.

Common mistake: People often talk of things *weighing*, say, a kilogram when really they mean the *mass* is 1 kilogram. A 1 kg bag of sugar, for example, has a mass of 1 kg and a weight of $1 \times g = 9.8$ N.

Worked examples

Example 1

A sledge of mass 12 kg is being pulled by a horizontal rope against a resistance of 2 N. The tension in the rope is 5 N. What is the acceleration of the sledge?

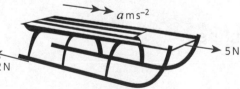

It is good to simplify a diagram by drawing the sledge as a block with arrows for the forces horizontal on the page.

Solution

The resultant horizontal force = (5 – 2)N and m = 12

So by Newton's 2nd law: 5 – 2 = 12a

The acceleration is 3 ÷ 12 = 0.25ms⁻².

Resultant force = *ma*.

Example 2

When a force of 0.5N is applied to a toy truck it travels 2.5m in 2s. Calculate the mass of the truck assuming it starts from rest and accelerates uniformly and there is no resistance to its motion.

The question has a mixture of information about forces as well as distances etc. Acceleration is the quantity that links them.

Solution

You can calculate the mass of the truck using *F* = *ma* if you first find the acceleration. You know *u* = 0, *s* = 1.5 and *t* = 2, so use:

$$s = ut - \tfrac{1}{2}at^2$$

Choose the equation that does not involve *v*.

giving $2.5 = 0 + \tfrac{1}{2}a \times 2^2$

$2.5 = 2a$

$a = 1.25$

Now use $F = ma$

$0.5 = m \times 1.25$

giving $m = 0.5 \div 1.25$

The mass of the truck is 0.4kg

Example 3

A racing driver of mass 70kg survived after hitting a wall at 48ms⁻¹ and stopping in 0.66m. What was the average magnitude of the force acting on the driver?

Solution

You are given *u* = 48, *v* = 0 and *s* = 0.66 so the acceleration can be calculated using:

$$v^2 = u^2 + 2as$$

Choose the equation that does not involve *t*.

$0 = 48^2 + 2 \times 0.66 \times a$

$-1.32a = 2304$

$a = -2304 \div 1.32$

$= -1745.45$

The car slowed down so the acceleration was negative

By Newton's 2nd law, the force acting was:

mass × acceleration = 70 × 1745.45N
= 122181.82 N or 122.2kN.

This is a massive force, but it is based on a true story.

Example 4

A girl of mass 50 kg is going up in a lift. Calculate the force between the girl and the floor of the lift:

 i   when it is accelerating upwards at $0.5\,\text{m s}^{-2}$

 ii when it is moving upwards at a steady speed

 iii when it is slowing down at $0.4\,\text{m s}^{-2}$.

Solution

 i   The figure shows the acceleration and forces acting on the girl. Notice that upwards is taken as the positive direction.

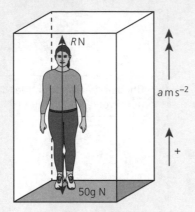

The girl's weight is 50gN and the acceleration, a = + 0.5.
The resultant force upwards is: R – 50gN.
By Newton's 2nd law:

$$R - 50g = 50 \times 0.5$$
$$R = 25 + 50 \times 9.8 = 515$$

The force between the girl and the floor is 515N.

 ii When the speed is steady, a = 0 and the forces on the girl are in equilibrium.

Hence the reaction force = the girl's weight
= 50gN (= 490N)

 iii When the lift is slowing down, a = –0.4 so

$$R - 50g = 50 \times (-0.4)$$
$$R = 50 \times 9.8 - 20 = 470$$

The force between the girl and the floor is 470N.

> Be clear that the acceleration is positive here.

> The value of R has changed. This could still be written $R - 50g = 0$

> The value of R is different again, but the equation has the same form.

Example 5

A skydiver of mass 80 kg dives from rest and reaches a speed of $96\,\text{m s}^{-1}$ in 20 s.

 i   What constant force would produce the same result?

 ii He then spreads himself out to reduce speed and after falling a further 220 m his speed is $53\,\text{m s}^{-1}$ (when he opens his parachute). What constant air resistance would give this result?

Solution

i Take the downwards direction to be positive, as in the figure on the right.

You are asked to find a constant force, so you can assume the acceleration is constant. Let the acceleration be a_1 ms^{-2}

$u = 0$, $v = 96$, $t = 20$, $a_1 =$?

Use $\qquad v = u + a_1 t$

$$96 = 0 + 20a_1$$

$$a_1 = 96 \div 20$$

The acceleration is 4.8 ms^{-2}

Now use $F = ma$ find the force

$$F = 80 \times 4.8$$

$$= 384$$

The force would be 384 N

ii Now $u = 96$, $v = 53$, $s = 220$, new acceleration $= a_2$ ms^{-2}

Use $\qquad\qquad v^2 = u^2 + 2a_2 s$

$$53^2 = 96^2 + 2 \times a_2 \times 220$$

$$53^2 - 96^2 = 440a_2$$

$$a_2 = -6407 \div 440$$

$$a_2 = -14.56...$$

The resultant force acting downwards is $80g - R$

So Newton's 2nd law gives:

$$80g - R = 80 \times (-14.56)$$

$$80 \times 9.8 + 80 \times 14.56 = R$$

$$R = 1948$$

The air resistance is 1.95 kN

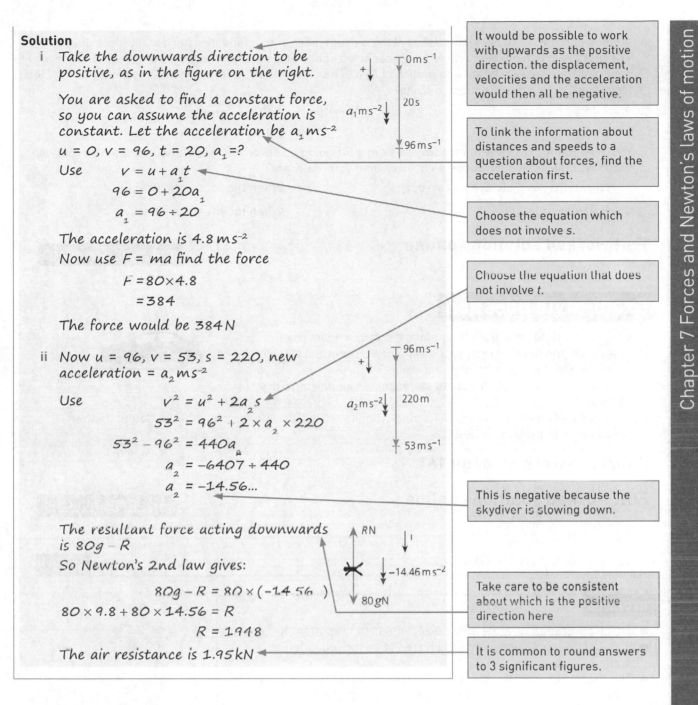

It would be possible to work with upwards as the positive direction. the displacement, velocities and the acceleration would then all be negative.

To link the information about distances and speeds to a question about forces, find the acceleration first.

Choose the equation which does not involve s.

Choose the equation that does not involve t.

This is negative because the skydiver is slowing down.

Take care to be consistent about which is the positive direction here

It is common to round answers to 3 significant figures.

Test yourself

1 In a road safety test a car containing a dummy of mass 60 kg is made to collide with a wall. Initially the car is moving at 16 ms^{-1} and it buckles when it hits the wall. What average force acts on the dummy if it moves forwards 0.8 m before stopping?

A 9600 N

B 9012 N

C 1200 N

D 588 N

2 A parachutist of mass 75 kg is descending with a downwards acceleration of 5.2 ms^{-2}. At this point the upthrust acting on the parachutist is U N. Later, the upthrust has doubled. What is his new acceleration?

A 2.6 ms^{-2}

B 1.67 ms^{-2}

C 0.6 ms^{-2}

D 0 ms^{-2}

3 A block of mass 12 kg is being pushed along a rough horizontal plane by a horizontal force of 24 N. The block starts from rest and when it reaches a speed of $1.5\,\text{m s}^{-1}$ the force is removed. There is a constant frictional resistance to motion of 15 N. The block slows down and comes to rest. For how long from start to finish is the block moving?

A 1.2 s

B 1.95 s

C 2 s

D 3.2 s

4 You are testing your new bathroom scales in a lift going up several floors without stopping in between. When will the scales show you as heavier than you really are?

A Never, they will always be correct.

B When the lift starts moving.

C Half way up.

D When the lift slows down.

Full worked solutions online

CHECKED ANSWERS

Exam-style question

A car of mass 1000 kg is travelling along a straight, level road.

i Calculate the acceleration of the car when a resultant force of 2000 N acts on it in the direction of its motion.

How long does it take the car to increase its speed from $5\,\text{m s}^{-1}$ to $12.5\,\text{m s}^{-1}$?

ii The car has an acceleration of $1.4\,\text{m s}^{-2}$ when there is a driving force of 2000 N. Calculate the resistance to motion of the car.

Short answers on page 161

Full worked solutions online

CHECKED ANSWERS

Connected particles

REVISED

> **Key facts**
>
> ● The acceleration is in the same direction as the resultant force.
> ● The equation obtained is often referred to as the 'equation of motion.'
> ● The magnitudes of the velocity and acceleration of two objects connected by a rod or a taut inextensible string are always the same.
> ● The tension in a string passing over a smooth light pulley is the same on both sides of the pulley.

When two or more objects are connected, there are usually several possible equations of motion you can write down.

You can consider all the particles together as one big mass, or you can treat one or more of them separately as separate masses each with their own forces.

Always remember that, when the connection is rigid (as is the case in these questions), the speed and acceleration of the objects must always be the same because they are attached.

When a pulley is involved, the directions might be different.

Worked examples

Example 1

Blocks A of mass 2 kg and B of mass 3 kg are joined together and sliding on smooth ice under the action of a force FN acting on the first block A. The tension in the joining rod is TN and the acceleration is a m s^{-2}. This is illustrated below.

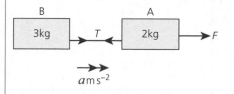

Write down the equations of motion for:

i the combined system

ii block A

iii block B.

Solution

i The first diagram to the right, shows the combined body as a large block with mass 5kg. The only force acting is FN. Its equation of motion is:

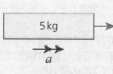

$$F = 5a \qquad (1)$$

ii The second diagram to the right, shows the block A. It has the forwards force FN and the backwards tension TN acting on it. Its equation of motion is:

$$F - T = 2a \qquad (2)$$

iii The third diagram to the right, shows the block B. It has only the forwards tension TN acting on it. Its equation of motion is:

$$T = 3a \qquad (3)$$

Notice that you can eliminate T from equations (2) and (3) by adding them:

$$F - T + T = 2a + 3a$$

Giving $F = 5a$ which is equation (1)

We can choose whether to think about one part of the system, or the other or the two parts combined together. The combined object has the tension as an internal force so need not be included in the equation.

Be clear that the block B does not experience the force F directly and so is not part of its equation of motion.

Hint: There are in fact only two equations rather than three which you can use.

Choose the two most appropriate ones. The third gives you no more information but can be used as a check.

Example 2

A locomotive of mass 80 000 kg is shunting a truck of mass 4000 kg into some sidings. The driving force of the locomotive is 2500 N and there are resistances of 200 N on each of the engine and the truck.

Calculate the force in the coupling between the locomotive and the truck.
Is it in tension or compression?

Solution

The diagram to the right shows the locomotive pushing the truck towards the right. For the time being, the force in the coupling is shown as a **tension** TN ($\rightarrow\leftarrow$). The acceleration is $a\,\mathrm{ms^{-2}}$

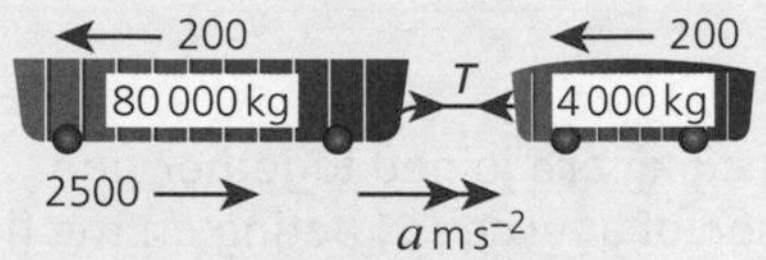

Forces in newtons

This is an arbitrary choice. Whichever you use, interpret the value of the answer at the end.

Using $F = ma$ for each separately gives:

Locomotive: $\qquad T + 2500 - 200 = 80000a$

$\qquad\qquad\qquad T + 2300 = 80000a \qquad$ (1)

Truck: $\qquad\qquad -T - 200 = 4000a \qquad$ (2)

Notice that the resultant force is taken to be in the same direction as the acceleration.

Eliminate a by multiplying (2) by 20 and substituting for 80 000a from (1):

$\qquad -20T - 4000 = 80000a$

So $\quad -20T - 4000 = T + 2300$

$\quad -4000 - 2300 = 21T$

$$T = -\frac{6300}{21} = -300$$

If the force in the coupling had been shown as a thrust, it would have turned out to be positive 300 N.

The **tension** is negative 300 N indicating that the force in the coupling is in fact a **thrust or compression** ($\leftarrow\rightarrow$) of 300 N. It acts forwards on the truck and backwards on the locomotive.

Example 3

A boat of mass 1500 kg is pulling a water skier of mass 65 kg. The driving force of the boat's engine is 8000 N. There are resistances of 1000 N on the boat and 100 N on the skier.

Calculate the acceleration and the tension in the towrope pulling the skier.

Solution

The diagram to the right shows that shows the horizontal forces acting on the boat and skier.

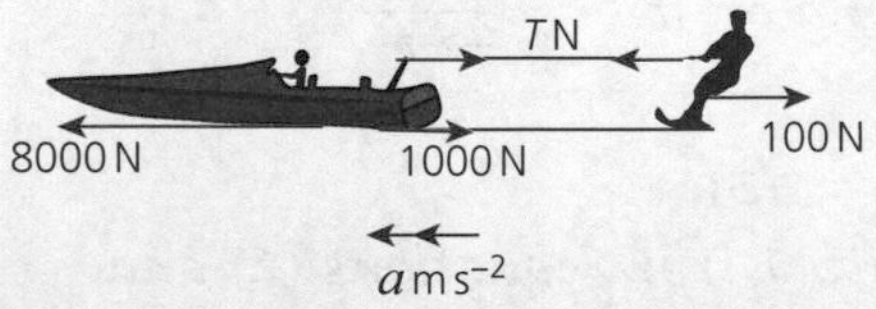

You can calculate the acceleration by considering them both together.

$$\left(8000 - 1000 - 100\right) = \left(1500 + 65\right) \times a$$

Resultant force

Total mass

$\qquad\qquad 6900 = 1565a$

$\qquad\qquad a = 6900 \div 1565$

$\qquad\qquad\quad = 4.41$

Acceleration

To calculate the tension, you need to consider the motion of the boat or the skier alone. For the water skier:

Notice that the boat and the water skier have the same acceleration.

$\qquad$ mass = 65 kg

$T - 100 = 65a$

$\qquad T = 65 \times 4.41 + 100$

$\qquad\quad = 386.6$

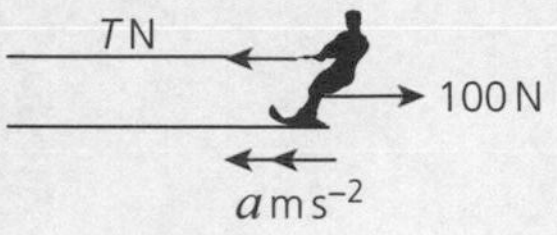

The tension in the rope is 387 N.

Example 4

Blocks of mass 0.5 kg (A) and 0.3 kg (B) are attached to the ends of a light string which hangs vertically over a smooth light pulley. The system is released from rest.

  i Draw a diagram to show the forces and acceleration of the blocks.

  ii Find the acceleration of the system.

  iii Find the tension in the string.

  iv Find the velocities of A and B after 2 s.

Solution

  i The forces acting are the weight of each block and the tension, T, in the string. The block A is heavier so it will move downwards and B will move upwards. The acceleration is $a\,\mathrm{ms^{-2}}$. The diagram below shows the forces and the accelerations of A and B.

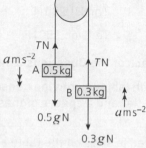

> The tension is the same either side of the pulley because the pulley is smooth.

> There are forces on the pulley as well but these are not required here.

  ii Write the equations of motion of A and B, each in the direction of its motion.

> Take the direction of motion of A and B to be the positive direction in each case. It does not matter that they are not the same direction.

For A ($\downarrow$): $0.5g - T = 0.5a$ (1)

For B ($\uparrow$): $T - 0.3g = 0.3a$ (2)

When these equations are added, the tension T is eliminated so:

$0.5g - 0.3g = 0.5a + 0.3a$

$0.2g = 0.8a$

The acceleration is $0.2g \div 0.8 = 2.45\,\mathrm{ms^{-2}}$.

  iii Substitute the value for the acceleration in (2) to calculate T:

> You can use either equation for this. You can use the other as well to check your answers.

$T - 0.3g = 0.3 \times 2.15$

$\quad T = 0.735 + 0.3 \times 9.8$

$\quad\quad = 3.675$

The tension is 3.675 N.

  iv Take the direction of motion to be positive.

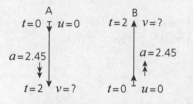

> Choose the equation that does not involve s.

For each block, $u = 0$, $a = 2.45$, $t = 2$

To calculate v after 2 seconds use

$v = u + at$

$v = 0 + 2.45 \times 2$

A has downwards velocity $4.9\,\mathrm{ms^{-1}}$

B is moving upwards at $4.9\,\mathrm{ms^{-1}}$

> It is important to include the direction of travel for each as the question asks for velocities and not speds.

Test yourself

Use the following information for Questions 1 and 2.

A car of mass 800 kg is pulling a trailer of mass 600 kg along a straight level road as shown in the diagram above. There is a resistance of 90 N on the car and 200 N on the trailer and they are slowing down at a rate of 0.8 m s^{-2}.

1 Calculate the braking force required.

   A 550 N   B 830 N

   C 1120 N   D 1410 N

2 By considering the motion of the trailer, calculate the force in the tow bar.

   A Thrust 680 N   B Thrust 280 N

   C Tension 200 N   D Tension 280 N

Use the following information for each of the Questions 3 and 4.

A block of mass 3 kg is held on a rough table. A light inextensible string attached to the block passes over a smooth pulley and a sphere of mass 2 kg hangs from the other end.

The block is then released and allowed to slide on the table against a friction force of 10 N. The tension in the string is T N and the acceleration of the system is a m s^{-2}.

3 Draw a diagram for yourself showing the forces acting on the block and the sphere and their accelerations. Use your diagram to write down their equations of motion.
Three of these equations are incorrect and one is correct. Which one is correct?

   A $T=2a$ B $T-10-3g=3a$

   C $2g-T=2ga$ D $T-10=3a$

4 Which of the following is the acceleration of the block?

   A 4.8 m s^{-2} B 1.92 m s^{-2}

   C 9.6 m s^{-2} D 29.6 m s^{-2}

Full worked solutions online

Exam-style question

A miniature train has an engine of mass 30 kg and four trucks each of mass 2.5 kg. The engine produces a driving force of 80 N and there are resistances of 48 N on the engine and 3 N on each truck.
Calculate:

i   The total resistance and the acceleration of the train.

ii The tension in the coupling between the last two trucks.

iii The tension in the coupling between the engine and the first truck.

Short answers on page 161

Full worked solutions online

Chapter 8 Variable acceleration

About this topic

You have already studied motion in a straight line where the acceleration is constant. You will now be using calculus to answer similar questions in the more general case where the acceleration need not be constant and may be expressed as a function of time.

Before you start, remember ...

- The language of kinematics and the basic definitions (covered in Chapter 6 of this guide).
- The use of position–time, velocity–time and acceleration–time graphs (in Chapter 6).
- Calculus technique such as:
  - ○ differentiation to find a gradient and stationary points
  - ○ integration to find the area under a graph.
- Curve sketching techniques.

> For regions above the t-axis the areas are taken to be + ve

> For regions below the t-axis the areas are taken to be − ve

- The area between a velocity–time curve and the t-axis represents displacement. This is often referred to as 'the area under the graph'. When v is negative the displacement is negative.
- Areas of the regions between an acceleration–time curve and the t-axis represent changes in velocity; when a is negative the change in velocity is negative.

Motion using calculus

REVISED ☐

Key facts

- In this section, the displacement s, the **instantaneous** velocity v and the **instantaneous** acceleration a of a particle moving in a straight line are all taken to be functions of time t.

displacement   velocity   acceleration

Differentiate →

$$s \qquad v = \frac{ds}{dt} \qquad a = \frac{dv}{dt} = \frac{d^2s}{d^2t}$$

← *Integrate*

$$s = \int v \, dt \qquad v = \int a \, dt \qquad a$$

> **Common mistake:** If you are given an expression for s, v or a in terms of t, only use the *constant acceleration* (*suvat*) formulae when you are sure that the acceleration is constant.

When you find velocity from acceleration or displacement from velocity, you use integration and so an arbitrary constant is involved. You need more information to find this constant but it is not needed if you use definite integration (i.e. integrate between limits).

Hint: Be sure that you know the language of kinematics. Especially important are the distinctions between: *displacement* and *distance travelled*; *velocity* and *speed*.

If in the slightest doubt, draw a sketch graph: position–time, velocity–time, distance–time etc.

Acceleration

Worked examples

Examples 1–5 are about a particle, P, that moves along the x-axis where the unit of length is the metre. Its velocity, v m s^{-1}, at time t seconds is given by $v = 12t^2 - 6t - 6$, where $-1 \leqslant t \leqslant 1$. When $t = -1$, $x = 3$.

Notice that all the units are S.I.

Example 1

Find an expression in terms of t for the acceleration, a m s^{-2}, of P at time t.

Solution

$$v = 12t^2 - 6t - 6$$

Differentiating, $a = \dfrac{dv}{dt} = 24t - 6$.

The *suvat* equations cannot be used here as the acceleration is not constant.

Example 2

i Draw the v–t graph for P for $-1 \leqslant t \leqslant 1$.

ii Find the greatest and least values of the velocity of P for $-1 \leqslant t \leqslant 1$.

You can sketch this by creating a table of values. Alternatively, you can see that there is a minimum point when $t = 0.25$ (from the derivative in Example 1) and the point where the quadratic graph crosses the axes.

Solution

i
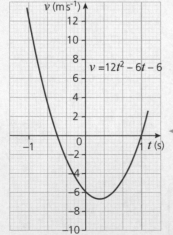

Greatest and least values in an interval are either at the ends or at a maximum or at a minimum point. The graph shows that the greatest value is at the left hand end-point and that the least value is inside the given interval $-1 \leqslant t \leqslant 1$.

ii The greatest value is when $t = -1$ so is given by $v = 12 \times (-1)^2 - 6 \times (-1) - 6 = 12$.
Thus the greatest velocity is 12 m s^{-1} (speed of 12 m s^{-1} in the direction Ox).
For the least value you need $a = 0$.
This occurs when $24t - 6 = 0$ and so $t = 0.25$.
When $t = 0.25$, $v = 12 \times 0.25^2 - 6 \times 0.25 - 6 = -6.75$
The least velocity is -6.75 m s^{-1} (speed 6.75 m s^{-1} in the opposite direction to Ox).

The least value of v occurs when $a = 0$. Since $a = \dfrac{dv}{dt}$ this is finding a stationary point.

The -ve sign shows that the motion is in the opposite direction to Ox

Position

Worked examples

Example 3

Find an expression for the position of P at time t.

Solution

$$x = \int \left(12t^2 - 6t - 6\right) dt$$

> x is used because the **position** is being found

$$= 12 \times \frac{t^{2+1}}{3} - 6 \times \frac{t^{1+1}}{2} - 6\frac{t^{0+1}}{1} + c$$
$$= 4t^3 - 3t^2 - 6t + c$$

> **Common mistake:** It is easy to forget the constant when integrating.

Now use the information that $x = 3$ when $t = -1$ to find c.

$$3 = 4 \times (-1)^3 - 3 \times (-1)^2 - 6 \times (-1) + c$$

> Remember that $x = 3$ when $t = -1$

so $3 = -4 - 3 + 6 + c$ and $c = 4$

> Put together the expression found above and the value for c to give a final answer.

Thus $x = 4t^3 - 3t^2 - 6t + 4$.

Example 4

What is the position of P when its velocity is instantaneously zero?

Solution

When $v = 0$, you have $12t^2 - 6t - 6 = 0$

> Use your calculator to solve this equation if you have this facility. Alternatively, you could use the quadratic formula or factorise to solve the equation.

$$6(2t^2 - t - 1) = 0$$
$$\Rightarrow 6(2t^2 - 2t + t - 1) = 0$$
$$\Rightarrow 6(2t + 1)(t - 1) = 0$$
$$\Rightarrow t = -0.5 \text{ or } t = 1.$$

> Look at the v–t graph in Example 2. When the curve crosses the t-axis, $v = 0$. This happens when $t = -0.5$ and $t = 1$.

You already know from Example 3 that the expression for the position of P is $x = 4t^3 - 3t^2 - 6t + 4$.

When $t = -0.5$, $x = 4 \times (-0.5)^3 - 3 \times (-0.5)^2 - 6 \times (-0.5) + 4$
$$= 5.75$$
When $t = 1$, $x = 4 \times 1^3 - 3 \times 1^2 - 6 \times 1 + 4 = -1$.

The positions are 5.75 m and –1 m.

Displacement

Displacement is the difference from one position to another. When you use the word you must also specify the starting position or time; for example, 'the displacement from point A' or 'the displacement from its position at time $t = -1$'. Displacement from the origin is the same as position.

Worked examples

Example 5

Find the displacement of P from its position when $t = -1$ to its position when $t = 1$.

Solution

You already know from Example 3 that the expression for the position of P is $x = 4t^3 - 3t^2 - 6t + 4$.

When $t = 1$: $x = 4 - 3 - 6 + 4 = -1$.

When $t = -1$: $x = -4 - 3 + 6 + 4 = 3$

So the displacement from its position when $t = -1$ to its position when $t = 1$ is $(-1) - (3) = -4$ m.

Examples 6–8 are about a particle, Q, that moves along the x-axis. The unit of length is the metre. Its velocity, v m s⁻¹, at time t seconds given by $v = 18t - 3t^2 - 24$ where $0 \leqslant t \leqslant 4.5$.

Example 6

  i   Draw the velocity–time graph for Q for $0 \leqslant t \leqslant 4.5$.
  ii Find an expression in terms of t for the displacement of Q from its position when $t = 1$.
  iii Use your answer to part **ii** to find the displacement of Q from its position when $t = 1$ to its position when $t = 3$.

Solution

  i

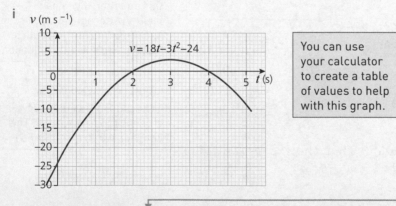

> You can use your calculator to create a table of values to help with this graph.

  ii Call s the displacement from Q, its position when $t = 1$.

When $t = 1$, the displacement is zero and so $s = 0$.

Now find $s = \int \left(18t - 3t^2 - 24\right) dt$, where $s = 0$ when $t = 1$.

Integrating, you get $s = 9t^2 - t^3 - 24t + c$.

Putting $s = 0$ when $t = 1$ gives $0 = 9 - 1 - 24 + c \Rightarrow c = 16$.

The required expression is $s = 9t^2 - t^3 - 24t + 16$.

  iii Now you just need to find the value of s when $t = 3$.

The required displacement is
$9 \times 3^2 - 3^3 - 24 \times 3 + 16 = 81 - 27 - 72 + 16 = -2$
so it is –2 m.

Displacements may be found directly as areas under a graph without finding positions first. Look at the v–t graph for P. This has been shaded to indicate the displacement from $t = -1$ to $t = 1$ with the region above the t-axis shown in black and that below it in red to emphasise that these represent positive (+) and negative (–) displacements, respectively. The overall displacement is the sum of the signed areas of these regions. The process of integration automatically gives the areas the correct signs.

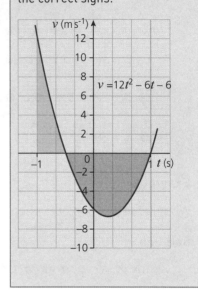

Notice that you do not have enough information to find an expression for the position.

Notice the change in notation, s for displacement, x for position.

The questions asks for the displacement from Q, which is the position of the particle when $t = 1$.

The negative (–) sign shows that the total displacement is in the opposite direction to Ox.

Hint: Look at Examples 5 and 6. In both you find the displacement from position at one time to position at another.

- In Example 5 you already had an expression for the position, x, at time t and so you used it to find the displacement

- In Example 6 you did not have an expression for the position so you first found one for the displacement from the position when $t = 1$ and then you used it.

- A further method, definite integration, is shown in Example 7. This is used when you have only been asked for the displacement from the position at one time to the position at a later time.

The 'best' method depends on what information you have and what expressions you have already found.

Hint: Always show how you evaluate an arbitrary constant. There will almost always be marks for doing this, even when its value is clearly zero.

Using definite integration

Worked example

Example 7

Find the displacement of Q from its position at $t = 1$ to its position at $t = 3$.

Solution

The displacement is given by

$$s = \int_{1}^{3} \left(18t - 3t^2 - 24\right) dt$$

$$= \left[9t^2 - t^3 - 24t\right]_{1}^{3}$$

$$= \left(9 \times (3)^2 - (3)^3 - 24 \times (3)\right) - \left(9 \times (1)^2 - (1)^3 - 24 \times (1)\right)$$

so $s = (-18) - (-16) = -2$ and the displacement is -2 m, as in Example 6.

The upper limit 3 is the end time. The lower limit 1 is the start time

It often useful to name the integrated function F(t). In this case $F(t) = 9t^2 - t^3 - 24t$. The value of the integral is expressed as $F(3) - F(1)$.

Distance travelled

In Example 7 you were asked to find the displacement of Q from $t = 1$ to $t = 3$. In Example 8 you will be asked to find the distance travelled between those times. To understand the difference, look at this diagram. It indicates displacements along the x-axis.

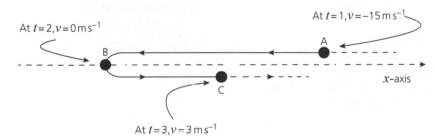

At $t = 2, v = 0\,\text{m s}^{-1}$

At $t = 1, v = -15\,\text{m s}^{-1}$

B

A

C

x-axis

At $t = 3, v = 3\,\text{m s}^{-1}$

You can infer the main features of the diagram from the v–t graph. (The graph seems to show that $v = 0$ when $t = 2$ and this may be confirmed by solving $v = 0$.)

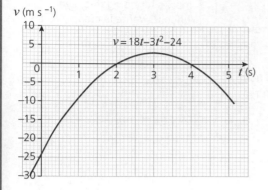

v (m s^{-1})

$v = 18t - 3t^2 - 24$

t (s)

- From $t = 1$ to $t = 2$, the velocity is in the negative direction and the displacement from A to B is negative.
- At $t = 2$, the velocity is zero.
- From $t = 2$ to $t = 3$, the velocity is in the positive direction and the displacement from B to C is positive.

The **displacement** is represented by the distance AC in the negative direction.

The **distance travelled** is the distance AB + the distance BC.

Worked example

Example 8

Find the distance travelled by Q from $t = 1$ to $t = 3$.

Solution

As you can see from the diagram above there are two parts to the journey, from A to B and from B to C. You first find the displacements for each part.

AB: $\displaystyle\int_{1}^{2}\left(18t - 3t^2 - 24\right)dt = \left[9t^2 - t^3 - 24t\right]_{1}^{2}$

$= (-20 - (-16)) = -20 + 16 = -4$

BC: $\displaystyle\int_{2}^{3}\left(18t - 3t^2 - 24\right)dt = \left[9t^2 - t^3 - 24t\right]_{2}^{3}$

$= (-18 - (-20)) = -18 + 20 = 2$

Travelling from A to B is a distance 4 m in the negative direction and ravelling from B to C is a distance 2 m in the positive direction, so the distance travelled is $4 + 2 = 6$ m.

> When distance is required, it is important to look separately at the parts of the journey in different directions.

> A displacement of -4 is a distance of 4 in the negative direction.

Hint: If you need displacement as well as distance travelled, it is easiest work the problem as in Example 8. Having found the displacement from A to B to be -4 m and the displacement from B to C to be 2 m, the **displacement of Q is** $(-4) + 2 = -2$ m; the **distance travelled** by Q is $|-4| + |2| = 4 + 2 = 6$ m.

> The symbol | | (called the modulus) means that the +ve value is taken

Test yourself

1 A toy is moving in a straight line and its velocity at time t seconds is $v\,\mathrm{m\,s^{-1}}$, where $v = -4t^2 + t + 5$ for $-1 \leqslant t \leqslant 3$. When is the acceleration of the toy zero?

A $t = 0$

B $t = 0.125$

C $t = -0.125$

D $t = -1$ and $t = 1.25$

2 A particle of grit, G, is stuck to the top of a piece of machinery that is moving up and down a vertical y-axis. The height of G above the ground is y metres at time ts where $y = 10t - 2t^2 - 8$. Determine the direction of motion and speed of G when $t = 3$.

A Downwards, $2\,\mathrm{m\,s^{-1}}$

B Downwards, $-2\,\mathrm{m\,s^{-1}}$

C Upwards, $2\,\mathrm{m\,s^{-1}}$

D Upwards, $4\,\mathrm{m\,s^{-1}}$

3 A particle is moving in a straight line and t seconds after passing through a point A its velocity is $V\,\mathrm{m\,s^{-1}}$, where $V = 4t - t^2 - 1$ for $0 \leqslant t \leqslant 5$. Draw a velocity–time graph before you start to answer the question. Use it to help you decide which one of the following contains only true statements.

A Initially the velocity of the particle is $-1\,\mathrm{m\,s^{-1}}$ and its acceleration is $4\,\mathrm{m\,s^{-2}}$ so the displacement of the particle from A after 3s is 15m.

B The initial velocity of the particle is $-1\,\mathrm{m\,s^{-1}}$ and its velocity after 3s is $2\,\mathrm{m\,s^{-1}}$ so the displacement of the particle from A after 3s is 1.5m.

C The speed of the particle after 1s is twice its initial speed; its greatest speed is $3\,\mathrm{m\,s^{-1}}$.

D The particle does not always travel in the same direction; its greatest velocity is $+3\,\mathrm{m\,s^{-1}}$.

In Questions 4 and 5, an insect is moving along an x-axis. At time t seconds, its velocity is $v\,\mathrm{m\,s^{-1}}$, where $v = 30t - 3t^2 - 63$.

4 Calculate the displacement of the insect from its position when $t = 2$ to its position when $t = 4$.

A $-2\,$m

B $2\,$m

C $-150\,$m

D $-76\,$m

5 Calculate the distance travelled by the insect in the time interval $2 \leqslant t \leqslant 4$.

A $157\,$m

B $-2\,$m

C $12\,$m

D $2\,$m

Full worked solutions online

Exam-style question

A particle moves along the x-axis with velocity $v\,\mathrm{m\,s^{-1}}$ at time ts given by $v = 18t - 12 - 6t^2$.

i Find an expression for the acceleration of the particle at time t.

ii Find the times t_1 and t_2, where $t_1 < t_2$, at which the particle has zero velocity.

iii Find the distance travelled between the times t_1 and t_2.

iv At time t_1 the particle passes through the point A. Does the particle pass through A on any later occasion? At time t_2 the particle passes through the point B. Does the particle pass through B on any later occasion?

v Find the distance travelled from $t = 0$ to $t = 3.5$.

Short answers on page 161

Full worked solutions online

Review questions (Mechanics)

1 A particle starts at the origin and moves to A which is 50 m due North in 8 s. It stays at A for 18 s and then travels 25 m due South to B at the same speed as before. Find

   i the displacement after both parts of the journey

   ii the total distance travelled

   iii the total time taken

   iv the average speed.

2 The motion of a particle is illustrated by the velocity–time graph. The positive direction is due North and the particle begins 200 m North of the origin.

   i Describe the motion of a particle.

   ii State the acceleration for each phase of the motion.

   iii Find the position of the particle after 40 s.

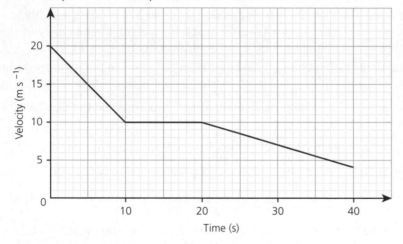

3 A space probe travels in a straight line. It accelerates uniformly from an initial velocity of $16 \, \text{m s}^{-1}$ in the negative direction to a velocity of $116 \, \text{m s}^{-1}$ in the positive direction in 120 s. It then continues to move along the same straight line with the same acceleration.
Find how long it takes the space probe to reach a point 60 km from its starting position. Give your answer in minutes and seconds to the nearest second.

4 Chunhua throws a ball vertically upwards from a point 1 m above the ground in his garden with a velocity of $25 \, \text{m s}^{-1}$. The garden is surrounded by a hedge that is 5 m high.
For how long is the ball visible from outside the garden?

5 A train consists of an engine of mass 8 tonnes and a carriage of mass 6 tonnes. It is travelling on level ground along a straight track at constant speed. The resistances to motion on the engine and the carriage are 2000 N and 1500 N respectively. The normal reactions between the engine and the track and the carriage and the track are N_1 and N_2. The engine exerts a driving force of D N. The tensions in the coupling is T N.

   i Draw a diagram showing all the forces acting on the engine and on the carriage.

   ii Find the values of D, N_1, N_2 and T.

6 A boy of mass 45 kg is in a lift which is ascending. Find the contact force between the boy and the lift floor in each case.

   i The lift is accelerating upwards at $2.5\,\mathrm{m\,s^{-2}}$.

   ii The lift is travelling at constant speed.

   iii The lift is slowing down with a deceleration of $3.5\,\mathrm{m\,s^{-2}}$.

7 A car of mass 1400 kg pulls a caravan on mass 1100 kg. It accelerates from rest to $10\,\mathrm{m\,s^{-1}}$ in 8 seconds.

   i Calculate the acceleration of the car and caravan assuming that it is constant.

   In an initial model, it is assumed that there is no resistance to motion.

   ii Find the driving force the car must provide according to this model.

   In a refined model it is assumed that the resistance is not negligible and that the driving force is 4025 N.

   iii Calculate the total resistance to motion according to this model.

   In this model it is also assumed that the resistance on the caravan is twice the resistance on the car.

   iv Calculate the tension in the tow-bar.

8 A sprinter starts running the 100m race from rest at the origin. Her velocity $v\,\mathrm{m\,s^{-1}}$ at time t seconds is given by

$$v = 4.8t - 0.6t^2 \quad \text{for } 0 \leqslant t \leqslant 4$$
$$v = 9.6 \quad \text{from } t = 4 \text{ until the end of the race.}$$

   i Find an expression for her acceleration at time t s

   Show that her acceleration zero when $t = 4$, and interpret this result.

   ii Find the distance she has run after

   a 4 s

   b 8 s.

   iii Calculate the time she takes to run 100 m.

Short answers on page 162

Full worked solutions online

CHECKED ANSWERS ☐

Exam preparation

Before your exam

- *Start revising early* – half an hour a day for 6 months is better than cramming in all-nighters in the week before the exam. Little and often is the key.
- *Don't procrastinate* – you won't feel more like revising tomorrow than you do today!
- Put your phone on *silent* while you revise – don't get distracted by a constant stream of messages from your friends.
- Make sure your *notes are in order* and nothing is missing.
- *Be productive* – don't waste time colouring endless revision timetables. Make sure your study time is actually spent revising!
- Don't just read about a topic. *Maths is an active subject* – you improve by answering questions and actually *doing* maths, not just reading about it.
- Time yourself on exam-style questions and check you are not spending too long on each question.
- Answer all the questions on as many practice papers as you can.
- Try *teaching a friend* a topic – teaching something is the best way to learn it yourself – that's why your teachers know so much!

The exam papers ...

You must take both Components 01 and 02 to be awarded the OCR AS Level in Mathematics B(MEI).

Component	Title	No. of marks		
01	**Pure Mathematics and Mechanics**	70	1 hour 30 minutes Written paper	**50% of total** AS Level
02	**Pure Mathematics and Statistics**	70	1 hour 30 minutes Written paper	**50% of total** AS Level

The content of this book covers all the Applied Mathematics content for both components.

Make sure you know these formulae for your exam...

From GCSE Maths you should know ...

Topic	Formula
Circle	$Area = \pi r^2$ $Circumference = 2\pi r$, where r is the radius
Parallelogram	$Area = base \times vertical\ height$
Trapezium	$Area = \frac{1}{2}h(a+b)$
Triangle	$Area = \frac{1}{2}\ base \times vertical\ height$
Prism	$Volume = area\ of\ cross\ section \times length$
Cylinder	$Volume = \pi r^2 h$ $Area\ of\ curved\ surface = 2\pi rh$ $Total\ surface\ area = 2\pi rh + 2\pi r^2$, where r is the radius and h is the height
Pythagoras' theorem	$a^2 + b^2 = c^2$
Trigonometry	$\cos\theta = \dfrac{adjacent}{hypotenuse}$ $\sin\theta = \dfrac{opposite}{hypotenuse}$ $\tan\theta = \dfrac{opposite}{adjacent}$

Topic	Formula	
Circle theorems	The angle in a semi-circle is a right-angle.	
	The perpendicular from the centre of a circle to a chord bisects the chord.	
	The tangent to a circle at a point is perpendicular to the radius through that point.	
Speed	$speed = \dfrac{distance}{time}$	
Distance–time graphs and velocity–time graphs	How to interpret them	

From AS Pure Maths you should know ...

REVISED ☐

Topic	Formula
Laws of indices	$a^m \times a^n = a^{m+n}$ $\dfrac{a^m}{a^n} = a^{m-n}$ $(a^m)^n = a^{mn}$ $a^{-n} = \dfrac{1}{a^n}$ $\sqrt[n]{a} = a^{\frac{1}{n}}$ $\sqrt[n]{a^m} = a^{\frac{m}{n}}$ $a^0 = 1$
Quadratic equations	The quadratic equation $ax^2 + bx + c = 0$ has roots $x = \dfrac{-b \pm \sqrt{b^2 - 4ac}}{2a}$

Topic	Formula
Coordinate geometry	• For two points (x_1, y_1) and (x_2, y_2): $\text{Gradient} = \dfrac{y_2 - y_1}{x_2 - x_1}$ $\text{Length} = \sqrt{(x_2 - x_1)^2 + (y_2 - y_1)^2}$ $\text{Midpoint} = \left(\dfrac{x_1 + x_2}{2}, \dfrac{y_1 + y_2}{2} \right)$. • The equation of a straight line with gradient m and y-intercept $(0, c)$ is $y = mx + c$. • The equation of a straight line with gradient m and passing through (x_1, y_1) is $y - y_1 = m(x - x_1)$. • Parallel lines have the same gradient. • For two perpendicular lines $m_1 m_2 = -1$. • The equation of a circle, centre (a, b) and radius r is $(x - a)^2 + (y - b)^2 = r^2$.
Trigonometry	For any triangle ABC *Area:* $\text{Area} = \dfrac{1}{2} ab \sin C$ *Sine rule:* $\dfrac{a}{\sin A} = \dfrac{b}{\sin B} = \dfrac{c}{\sin C}$ or $\quad \dfrac{\sin A}{a} = \dfrac{\sin B}{b} = \dfrac{\sin C}{c}$ *Cosine rule:* $a^2 = b^2 + c^2 - 2bc \cos A$ or $\quad \cos A = \dfrac{b^2 + c^2 - a^2}{2bc}$ *Identities:* $\sin^2 \theta + \cos^2 \theta \equiv 1$ $\tan \theta = \dfrac{\sin \theta}{\cos \theta}, \quad \cos \theta \neq 0$
Transformations	$y = f(x + a)$ is a translation of $y = f(x)$ by $\begin{pmatrix} -a \\ 0 \end{pmatrix}$. $y = f(x) + b$ is a translation of $y = f(x)$ by $\begin{pmatrix} 0 \\ b \end{pmatrix}$. $y = f(ax)$ is a one-way stretch of $y = f(x)$, parallel to x-axis, scale factor $\dfrac{1}{a}$. $y = af(x)$ is a one-way stretch of $y = f(x)$, parallel to y-axis, scale factor a. $y = f(-x)$ is a reflection of $y = f(x)$ in the y-axis. $y = -f(x)$ is a reflection of $y = f(x)$ in the x-axis.
Polynomials and binomial expansions	*The Factor theorem:* If $(x - a)$ is a factor of $f(x)$ then $f(a) = 0$ and $x = a$ is a root of the equation $f(x) = 0$. Conversely, if $f(a) = 0$ then $(x - a)$ is a factor of $f(x)$. *Pascal's triangle:* $\begin{array}{ccccccccc} & & & & 1 & & & & \\ & & & 1 & & 1 & & & \\ & & 1 & & 2 & & 1 & & \\ & 1 & & 3 & & 3 & & 1 & \\ 1 & & 4 & & 6 & & 4 & & 1 \end{array}$ *Factorials:* $n! = n \times (n - 1) \times (n - 2) \times \ldots \times 1$

Topic	Formula	
Differentiation	*Function*	*Derivative*
	$y = kx^n$	$\dfrac{dy}{dx} = knx^{n-1}$
	$y = e^{kx}$	$\dfrac{dy}{dx} = ke^{kx}$
	$y = f(x) + g(x)$	$\dfrac{dy}{dx} = f'(x) + g'(x)$
Integration	*Function:* $\displaystyle\int kx^n dx$	*Integral:* $\dfrac{kx^{n+1}}{n+1} + c$
Vectors	$\overrightarrow{AB} = \overrightarrow{OB} - \overrightarrow{OA}$	
	If $\mathbf{a} = x\mathbf{i} + y\mathbf{j}$ then $|\mathbf{a}| = \sqrt{x^2 + y^2}$	
Exponentials and logarithms	$y = \log_a x \Leftrightarrow a^y = x$ for $a > 0$ and $x > 0$	
	$\log xy = \log x + \log y \qquad \log \sqrt[n]{x} = \dfrac{1}{n}\log x \quad \log_a a = 1$	
	$\log \dfrac{x}{y} = \log x - \log y \qquad \log \dfrac{1}{x} = -\log x \quad e = 2.718...$	
	$\log x^n = n \log x \qquad\qquad \log 1 = 0 \qquad\quad \log_e x = \ln x$	

From AS Mechanics you should know …

REVISED ☐

Topic	Formula
Forces and equilibrium	Weight = mass, g
	Newton's 2nd law in the form $F = ma$
Calculus expressions	displacement $\qquad$ velocity $\qquad$ acceleration
	Differentiate →
	$s \qquad\qquad v = \dfrac{ds}{dt} \qquad\qquad a = \dfrac{dv}{dt} = \dfrac{d^2 s}{dt^2}$
	← *Integrate*
	$s = \displaystyle\int v\, dt \qquad\qquad v = \displaystyle\int a\, dt \qquad\qquad a$

From AS statistics you should know …

REVISED ☐

Topic	Formula
Sample mean	$\bar{x} = \dfrac{\sum x}{n} = \dfrac{\sum fx}{\sum f}$

Formulae that will be given

REVISED

Make sure you are familiar with the formula book you will use in the exam.

Here are the formulae you are given for the They don't get a different set of formulae for the applied maths section so this is misleading part of the exam.

> If you are studying A Level Mathematics you will be given a longer formula sheet. See the A level Pure Mathematics Revision Guide

Topic	Formula
Binomial series	$(a + b)^n = a^n + {}^nC_1 a^{n-1}b + {}^nC_2 a^{n-2}b^2 + \ldots + {}^nC_r a^{n-r}b^r + \ldots + b^n \ldots (n \in \mathbb{N})$, where ${}^nC_r = \begin{pmatrix} n \\ r \end{pmatrix} = \dfrac{n!}{r!(n-r)!}$ *This is on your AS formula sheet but you only need this for A level Maths.* $(1 + x)^n = 1 + nx + \dfrac{n(n-1)}{2!}x^2 + \ldots + \dfrac{n(n-1)\ldots(n-r+1)}{r!}x^r + \ldots (|x| < 1, n \in \mathbb{R}$
Differentiation from first principles	$f'(x) = \lim\limits_{h \to 0} \dfrac{f(x+h) - f(x)}{h}$
Sample variance	$s^2 = \dfrac{1}{n-1}S_{xx}$ where $S_{xx} = \sum(x_i - \overline{x})^2 = \sum x_i^2 - \dfrac{\left(\sum x_i\right)^2}{n} = \sum x_i^2 - n\overline{x}^2$
Standard deviation	$s = \sqrt{\text{variance}}$
The binomial distribution	If $X \sim B(n, p)$ then $P(X = r) = {}^nC_r p^r q^{n-r}$ where $q = 1 - p$ Mean of X is np
Kinematics	*Motion in a straight line* $v = u + at$ $s = ut + \frac{1}{2}at^2$ $s = \frac{1}{2}(u + v)t$ $v^2 = u^2 + 2as$ $s = vt - \frac{1}{2}at^2$

Make sure you know these formulae for your exam...

During your exam

Watch out for these key words:

- **Exact** … leave your answer as a simplified surd, fraction or power.

  Examples: $\ln 5$ ✓ 1.61 ✗ e^2 ✓ 7.39 ✗

  $2\sqrt{3}$ ✓ 3.26 ✗ $1\frac{5}{6}$ ✓ 1.83 ✗

- **Determine** … justification should be given for any results found, including working where appropriate.

- **Give/State/Write down** … no working is expected – unless it helps you.

  The marks are for the answer rather than the method.

  > **Example:** The equation of a circle is $(x + 2)^2 + (y - 3)^2 = 13$
  >
  > **Write down** the radius of the circle and the coordinates of its centre. ◄——

  > Make sure you give both answers!

- **Prove/Show that** … the answer has been given to you. You must show full working otherwise you will lose marks. Often you will need the answer to this part to answer the next part of the question. Most of the marks will be for the method.

  > **Example: i Prove** that $\sin x - \cos^2 x \equiv \sin^2 x + \sin x - 1$

- **Hence** … you **must** follow on from the given statement or previous part. Alternative methods may not earn marks. ◄——

  > **Example: ii Hence solve** $\sin x - \cos^2 x = -2$ for $0° \leqslant x \leqslant 180°$

  > Remember that if you couldn't answer part i you can still go on and answer part ii.

- **Hence or otherwise** … there may be several ways you can answer this question but it is likely that following on from the previous result will be the most efficient and straightforward method.

  > **Example:** Factorise $p(x) = 6x^2 + x - 2$
  >
  > Hence, or otherwise, solve $p(x) = 0$

- **You may use the result …** indicates a given result that you would not always be expected to know, but which may be useful in answering the question. This does not necessarily mean that you have to use the given result.

- **Plot …** you should mark points accurately on graph paper provided in the Printed Answer Booklet. You will either have been given the points or have had to calculate them. You may also need to join them with a curve or a straight line.

  > **Example:** Plot this additional point on the scatter diagram.

- **Draw …** Learners should draw to an accuracy appropriate to the problem. They are being asked to make a sensible judgement about the level of accuracy which is appropriate.

  > **Example:** Draw a diagram showing the forces acting on the particle.
  > Draw a line of best fit for the data.

- **Sketch (a graph) ...** You should draw a diagram, not necessarily to scale, showing the main features of a curve. These are likely to include at least some of the following:

  o Turning points
  o Asymptotes
  o Intersection with the y-axis
  o Intersection with the x-axis
  o Behaviour for large x (+ or −)

  Any other important features should also be shown.

  Example: Sketch the curve with equation $y = \dfrac{1}{(x-1)}$.

- **In this question you must show detailed reasoning ...** You must give a solution which leads to a conclusion showing a detailed and complete analytical method. Your solution should contain sufficient detail to allow the line of your argument to be followed. This does not prevent you from using of a calculator when tackling the question, e.g. for checking an answer or evaluating a function at a given point, but it is a restriction on what will be accepted as evidence of a complete method.

Watch out for these common mistakes:

- ✘ Miscopying your own work or misreading/miscopying the question
- ✘ Not giving your answer as coordinates when it should be. E.g. when finding where two curves meet
- ✘ Check the units in the question e.g. length given in cm not m
- ✘ Giving your answer as coordinates inappropriately e.g. writing a vector as coordinates
- ✘ Not finding y coordinates when asked to find coordinates
- ✘ Not finding where the curve cuts the x **and** y axes when sketching a curve
- ✘ Using a ruler to draw curves
- ✘ Not using a ruler to draw straight lines
- ✘ Spending too long drawing graphs when a sketch will do
- ✘ Not stating the equations of asymptotes when sketching an exponential or reciprocal curve
- ✘ Not simplifying your answer sufficiently
- ✘ Rounding answers that should be **exact**
- ✘ Rounding errors – don't round until you reach your final answer
- ✘ Giving answers to the wrong degree of accuracy - use 3 s.f. unless the questions says otherwise
- ✘ Not showing any or not showing enough working – especially in 'show that' or 'proof' questions
- ✘ Showing full working for a mean or standard deviation calculation when you were expected to write down an answer using your calculator functions
- ✘ Giving two contradictory answers – you need to make it clear which your answer is if you change your mind

Once you have answered a question **re-read the question** making sure you've answered **all** of it. It is easy to miss the last little bit of a question.

Check your answer is ...

✓ to the correct accuracy
✓ a sensible size in the context of the question
✓ in the right form
✓ complete ... have you answered the whole question?

If you do get stuck ...

Keep calm and don't panic.

✓ **Reread the question** ... have you skipped over a key piece of information that would help? Highlight any numbers or key words.

✓ **Draw** ... a diagram. This often helps. Especially in questions on graphs, coordinate geometry and vectors, a sketch can help you see the way forward.

✓ **Look** ... for how you can re-enter the question. Not being able to answer part **i** doesn't mean you won't be able to do part **ii**. Remember the last part of a question is not necessarily harder.

✓ **Move on** ... move onto the next question or part question. Don't waste time being stuck for ages on one question, especially if it is only worth one or two marks.

✓ **Return later** ... in the exam to the question you are stuck on – you'll be surprised how often inspiration will strike!

✓ **Think positive!** You are well prepared, believe in yourself!

Good luck!

Answers

Here you will find the answers to the Target your revision exercises, Exam-style questions and Review questions in the book. Full worked solutions for all of these are available online at **www.hoddereducation.co.uk/myrevisionnotesdownloads**. Note that answers for the Test yourself multiple choice question are available online only.

STATISTICS

Target your revision (page 1)

1 sample, population, sample, population

2 Number list of all students.

   Choose 40 different random integers in range 1 to 900.

   Find the students with those numbers.

3 Opportunity sample

4 The method only samples people who are able to get to the centre during its current opening times.

5 Lack of sampling frame.

6 Categorical

7 The graph shows average earnings for different age groups at one time; it does not show how earnings for particular women change with age

8 64%

9 There are many possible answers such as
   ● Boys in the first year of life had the greatest chance of dying.
   ● Other than the first year of life, men in their twenties were the most likely to die.

   See the full solution online for other answers.

10 Answer in range 350 000 and 400 000.

11 Increase in percentage employed in services.

   Decrease in percentage employed in manufacturing.

12 The times are similar.

13 There are several possible answers including the following
   ● House prices in Romford have a greater spread than house prices in Nelson.
   ● House prices in Romford tend to be higher than in Nelson.
   ● Over half the houses sold in Nelson were cheaper than any house sold in Romford.

   See the full solution online for other answers.

14 Answer in range 46 to 48.

15 Bimodal

16 i Positive correlation.

    ii No. One possible reason is that a curve would fit the data better. See the online solution for other possible reasons.

17 The point close to (0, 180) is an outlier. Zero blood pressure is impossible so it almost certain to be the result of an error.

18 i Too many sectors in the pie chart make it hard to read.

    ii There are many possible answers, for example to draw a bar chart instead. See the online solution for other answers.

19 i Male £36 648 (nearest £)

    ii Female £29 031 (nearest £)

    iii The males have a higher average salary.

20 i Median is 37 s. Interquartile range is 17 s.

    ii Median is 36 s. Interquartile range is 22 s.

    iii Both groups have nearly the same average but Group 2 has a higher spread.

21 i Male £13 663 (nearest £)

    ii Female £8445 (nearest £)

    iii Male salaries are more spread out than female salaries.

22 There is at least one high outlier but there are no low outliers.

23 0.0144

24 $\frac{4}{9}$

25 The binomial model is not suitable. The probability of 'success' should be the same each time but in this case it is not.

26 0.057

27 100

28 0.1

29 If the null hypothesis is true, the probability of it being rejected is 5%.

30 $p \neq 0.55$

31 $H_0 : p = 0.9$

   $H_1 : p < 0.9$

   where p is the proportion of callers who wait less than 5 minutes.

32 $X \leq 4$ and $X \geq 17$.

33 There is sufficient evidence, at the 5% level, that there has been a decrease in the proportion of faulty plates.

Data collection (page 11)

i   Sixth form students in Leeds.

ii Get a list of all sixth form students in Leeds and number each one. Choose 100 different random numbers in the range 1 to total number of students. Choose the students with those random numbers.

iii Opportunity sample.

iv Two distinct sources of bias. Some possible examples are given below.
   ● The students at that sixth form college may not be representative of the population.
   ● Students who take politics may not be representative of the population.
   ● Seeing when other students put up their hands may lead to students not answering honestly.

Intepreting graphs for one variable (page 19)

i   Start the vertical axis at zero.

   Have a key to say which bars represent males and which represent females.

ii For both males and females life expectancy has gone up.

   The difference between male and female life expectancy is decreasing.

   Other possible answers include: Male life expectancy is lower than female life expectancy.

iii For a small area, the deaths in one year might not be representative; looking at several years increases the sample size and increases the representativeness.

iv The two life expectancies from the graph are about 72 and 80. There are roughly equal numbers of males and females at birth so the average of these would be a reasonable life expectancy for all people.

   $\frac{72+80}{2}=76$

Averages (page 25)

i   19 is not correct; this is half way between 17 and 21 but the oldest person in the group could be nearly 22 so the midpoint is $\frac{17+22}{2}=19.5$.

ii Mean age is 24.4 years (to 1 d.p.).

iii The answer is only an estimate because the data were grouped. The exact ages are not known and so the midpoint of each group has been used to estimate them.

iv 25.0

Diagrams for grouped data (page 34)

i   For the blue graph, in the 50 to 60 group there are $10-2=8$ people.

   [Answer in range 6 to 12 with correct working]

   In the histogram, there are $10 \times 2 = 20$ people.

   The red graph must be the one representing males.

ii The total frequency for the blue graph is slightly higher than the total frequency for the red graph so there were more females than males but the numbers were fairly similar.

iii A   From the histogram, it is $0.2\times10=2$.

   B $91-79=12$ [Answer in range 10 to 14 with correct working]

iv Female pulse rates tend to be higher than male pulse rates.

Median and quartiles (page 43)

i   Median$=4.35$ cm

   $IQR=4.85-3.75=1.1$ cm

ii $1.5\times IQR=1.5\times1.1=1.65$

   $4.85+1.65=6.5$ cm

   $3.75-1.65=2.1$ cm

   Leaves shorter than 2.1 cm, or longer than 6.5 cm could be considered outliers.

iii Up to 5 possible lower end outliers.

   Up to 3 possible upper end outliers but you cannot tell whether there were any leaves longer than 6.5 cm or shorter than 2.1 cm without seeing the original data. So Alex was wrong to use the word 'must'.

Standard deviation (page 48)

i   The sample mean is 410.47 g.

   The sample standard deviation is 12.93 g (2 d.p.).

ii The sample mean is 393.57 g.

   The sample standard deviation is 74.03 g (2 d.p.)

iii The loaves of bread in the second sample are slightly lighter, on average.

   Each sample mean is fairly close to the intended weight of 400 g.

   The weights of the second sample are much more spread than those of the first.

   It looks as if there is something wrong with the process when the second sample is taken.

Bivariate data (page 56)

i   Positive correlation.

ii Region where both are healthy is shaded.

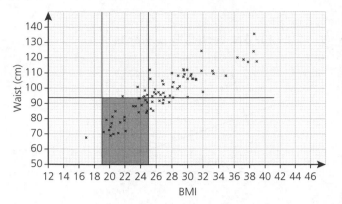

iii Slightly less than half of them have a healthy BMI.

$$\frac{36}{89} = 40.4...\% \approx 40\%$$

Working with probability (page 64)

i 12

ii 0.913 (to 3 s.f.)

iii P(no fakes) < 0.5

23 coins.

iv E.g. in a part of the country where there are more fakes than usual

Probability distributions (page 69)

i $p = \frac{11}{30}$

ii If $X = 4$, then the fifth girl would also get the gift she had bought, so X would be 5, not 4.

iii $P(X=5) = \frac{1}{5} \times \frac{1}{4} \times \frac{1}{3} \times \frac{1}{2} \times 1$

$P(X=5) = \frac{1}{120}$

Introducing the binomial distribution (page 75)

i $P(\text{double } 6) = \frac{1}{6} \times \frac{1}{6} = \frac{1}{36}$

ii $P(X=1) = {}_{24}C_1 \times \left(\frac{35}{36}\right)^{23} \times \left(\frac{1}{36}\right)^1 = 0.349$ (3 s.f.)

iii P(at least one double 6) $= 1 - 0.508\,596..... = 0.491\,40...$
$= 0.491$ (3 s.f.)

Cumulative probabilities and expectation (page 81)

Note that answer i can be found online.

ii 6

iii 0.0730

Introducing hypothesis testing using the binomial distribution (1-tail tests) (page 89)

i $H_0 : p = 0.75$

$H_1 : p < 0.75$

where p = the proportion of people in the population that the treatment is effective for.

ii There is insufficient evidence, at the 5% level, to suggest that the researcher is exaggerating.

iii The assumption that people respond to the treatment independently is needed so that a binomial distribution can be used. Responses are likely to be independent as long as people do not find out about how others are responding.

2-tail tests (page 93)

i $H_0 : p = 0.15$

$H_1 : p \neq 0.15$

where p is the proportion of young people in the area who are trying to give up smoking.

ii The critical region is $X \geq 7$.

iii There is no evidence, at the 5% level, that the proportion of young people in the area who are trying to give up smoking is any different to 15%.

Review questions (Statistics) (page 94)

1 i Cluster

ii 50.8 (to 1 d.p.)

iii Suitable reason, e.g. people may not remember exactly how many lessons they had so asking them to choose a group is realistic accuracy

Other possible reasons are given in the online answers.

2 i Two distinct comments, e.g.
- The distributions have similar shape and location.
- The data for 1896–1905 are more spread out than those for 1906–2005.

Further possible comments are given in the online answers.

ii Median 5.1°C; *IQR* 1.65°C

iii The temperatures in 1996–2005 were a little higher and less spread out but it would be hard to cite them as evidence for global warming.

OR

The data are for one place at two times so it is not possible say whether this is part of a worldwide change over time.

3 i A £51 587 to nearest pound

B £39 600

C £63 800

   ii The new trustee earns less than all the others. Including her will bring down the mean as the new trustee only adds a small amount to the total but this total now needs to be shared between 10 people so the mean is reduced.

    For the median, the new number will be at the bottom of the list and there will be 10 numbers in the list instead of 9. New median is £35 340. This is lower than before as there are now two numbers in the middle.

    The midrange will be lower because you will be finding halfway between 110 000 and a lower number than before.

   iii Median because it is not affected by a very high or low value.

4   0.65

5 $\dfrac{7}{216}$

6 **i**   0.0182 (3 s.f.)

   ii There is evidence, at the 5% level, to suggest that the probability of getting a four is less than $\dfrac{1}{8}$.

7 **i** $H_0 : p = 0.08$

      $H_1 : p \neq 0.08$

     where p is the proportion of vegetarians in the adult population in the area.

   ii $\{0, 1, 2\} \cup \{x \geq 15\}$ where x is the number of vegetarians in the sample.

MECHANICS

Target your revision (page 96)

1 **i**   50 s

   ii

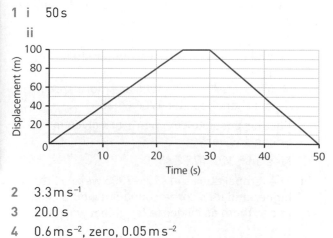

2   3.3 m s^{-1}

3   20.0 s

4   0.6 m s^{-2}, zero, 0.05 m s^{-2}

5 **i**   −1.25 m s^{-2}

   ii 40 m

6   4 s

7 **i**   0.495 s

   ii 4.85 m s^{-1}

8   12.5 m s^{-1}

9 $11g = 107.8$ N, vertically upwards

10

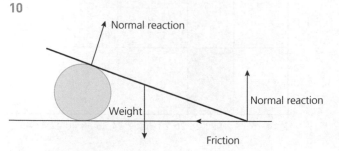

11 The frictional force is mg

12 $x = -3$, $y = 34$

13 **i**   490 000 N

    ii   6.2 m s^{-2}

14 6.06 N

15 From left to right $T_1 - 0.5g = 0.5a$, $T_2 - T_1 = 2a$, $1.5g - T_2 = 1.5a$

16 a is not constant

17 Yes, when $t = 16$ s

18 6 m

Motion (page 104)

i **A**   10 m s^{-1}

   B   300 s, 7 m s^{-1}

   C   −0.125 m s^{-2}

ii **A**

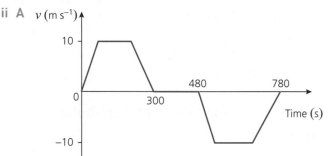

Using graphs (page 110)

i

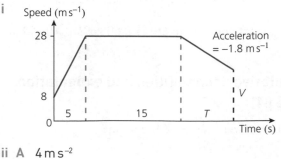

ii **A**   4 m s^{-2}

    B   90 m

iii $T = 10$, $V = 10$

iv A 23.3 m s⁻¹

B

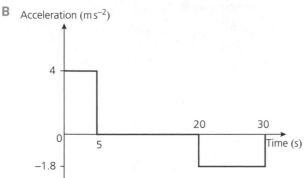

Using the constant acceleration formula (page 114)

i 38.5 m

ii 8.5 s

Vertical motion under gravity (page 119)

i 22.5 m

ii 11.5 m

Forces (page 125)

i

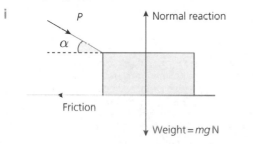

ii

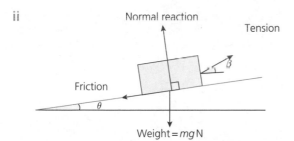

Newton's 1st law of motion (page 130)

i

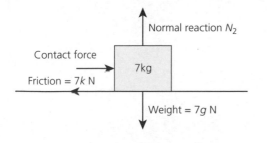

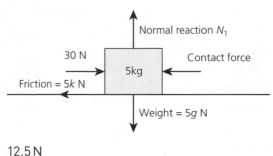

ii 12.5 N

iii 68.6 N

iv 17.5 N

Applying Newton's 2nd law along a line (page 134)

i 2 m s⁻², 3.75 s

ii 600 N

Connected particles (page 138)

i 60 N, 0.5 m s⁻²

ii 4.25 N

iii 17 N

Motion using calculus (page 145)

i $a = 18 - 12t$

ii $t_1 = 1$, $t_2 = 2$

iii 1 m

iv The particle will pass again through A but not pass again through B.

v 19.5 m

Review questions (Mechanics) (page 146)

1 i 25 m

ii 75 m

iii 30 s

iv 2.5 s⁻²

2 i The particle is initially travelling North with a velocity of 20 m s⁻¹. It decelerates at a constant rate to 10 m s⁻¹ after 10 s. It then travels at constant speed for the next 10 s. Then it decelerates, again at a constant rate, reaching 4 m s⁻¹ after another 20 s.

ii −1 m s⁻², 0 m s⁻², −0.3 m s⁻²

iii 590 m North of the origin.

3 6 minutes 45 seconds

4 4.8 s

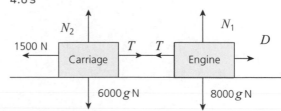

5 $D = 3500$ N, $N_1 = 78400$ N, $N_2 = 58800$ N, $T = 1500$ N

6 i 553.5 N

ii 441 N

iii 283.5 N

7 i 1.25 m s^{-2}

ii 3125 N

iii 900 N

iv 1975 N

8 i $a = 4.8 - 1.2t$. At $t = 4$ she has reached her maximum velocity.

ii A 25.6 m

B 64 m

iii 11.75 s